Parkinson's Warrior: DBS

Parkinson's Warrior: DBS

By Nick Pernisco

Published by Connected Neurosciences LLC

Connected Neurosciences LLC
1037 NE 65th St. #80832
Seattle, WA 98115

www.ParkinsonsWarrior.com
info@parkinsonswarrior.com

Ordering Information:

Quantity sales. Special discounts are available on quantity purchases by corporations, associations, and others. For details, contact the publisher at the address above.

Orders by U.S. trade bookstores and wholesalers. Please contact Ingram Book Company: Tel: (800) 937-8200; Email: customerservice@ingrambook.com.

Printed in the United States of America

Cover design by Marianna Bernshteyn

Ebook: ISBN 9781087886794
Print: ISBN 9781087886787

First Edition

10 9 8 7 6 5 4 3 2 1

Parkinson's Warrior: Deep Brain Stimulation is dedicated to my wife, Rosaline Bernstein, to my brother, Cesar Pernisco, and to my mother, Teresa Pernisco.

Preface

First things first – I'm not a doctor nor a lawyer, I don't pretend to be a doctor or a lawyer, and I don't play one on TV (or YouTube). What I *am* is a well-informed, engaged Person with Parkinson's (PWP). My intent is to provide you with information and inspiration to help guide you through your Parkinson's journey. While you will learn all about Deep Brain Stimulation, an advanced treatment for Parkinson's, and you will be able to speak the "lingo" intelligently with your doctors, your family, and your friends, this is not a medical book, per se. This is one Person with Parkinson's reaching out to hold the hand of another Person with Parkinson's.

Always seek your doctor's advice and follow their instructions before starting any new medication, beginning any exercise routine, changing anything about your routine, or plowing ahead with having DBS surgery. In these pages, I will share my experiences of having Deep Brain Stimulation surgery, and give you a lot to think about as you consider this therapy for yourself. My hope is that you will learn more about Deep Brain Stimulation and discover how it might change your life. I hope you take this information, do more research on your own, then be informed enough to have a true discussion with your medical team – one in which there is a back and forth of ideas, not one in which your doctor says "this is how it will be because I'm the doctor." A good doctor will always welcome your input, and a great doctor will admit when they don't know enough about a topic and will be open to continue learning.

Ultimately, the decision is up to you. But as every Parkinson's Warrior knows, information is power. The more of it you have, the better off you will be. This book is meant to inform, educate, and inspire. After reading this book, you will know the good and the not-so-good about Deep Brain Stimulation surgery. After reading it, take to the internet with questions, speak with doctors, consult with your local Parkinson's organizations, speak with others who have had the surgery, and only then decide if this surgery is right for you.

I would also like to say up front that I am NOT being sponsored by any company or organization. The only money I make is through book sales, and perhaps through a few sales of my app, Parkinson's LifeKit, which I mention throughout the book because it has helped me. DBS has had a profound impact on my life, but I will not recommend a specific company, and I will readily indicate any faults I see in any product. By buying this book, we have entered into a sacred bond between reader and writer. I am looking to inform, educate, and inspire, not get rich.

Acknowledgements

This book could not have been written without the help of so many people. To everyone in the Seattle Parkinson's community, including my friends at the APDA, thank you for your inspiration. To everyone at the Amsterdam Medical Centre, thank you for your dedication and kindness. To my friends and family, thank you for being there and for your love.

Introduction

On December 6th, 2018 I underwent Deep Brain Stimulation surgery to alleviate the symptoms of Parkinson's Disease. My first symptoms had appeared in 2009, so it was almost exactly nine years later that I would go under the drill and knife to get some relief. I had some apprehension at first – I mean, it's brain surgery – but I'm glad I went through with it. Although DBS was not a cure for Parkinson's, the surgery set back the clock on my symptoms about 10-12 years.

In this sequel to the original Parkinson's Warrior book, I detail the circumstances that lead me to consider Deep Brain Stimulation, the essential details about what the surgery does and how it works, and a first-hand account of my own surgery. Just like in the original Parkinson's Warrior book, I pull no punches. I give you every detail, no matter how frightening it might sound. My hope is that, after reading this book, Deep Brain Stimulation won't sound so frightening.

The reason I wrote this book was to demonstrate that an option exists to alleviate Parkinson's symptoms, and that people with Parkinson's do not have to lose hope when their symptoms worsen. I had Deep Brain Stimulation surgery to give me time to live my life more normally until a cure for Parkinson's is discovered.

The second part of this book involves my life after DBS surgery. I discuss some expectations I had about the surgery, how I felt after surgery, and how having a Warrior mentality helped me adjust to a new normal. I hope this second section will inspire you and make you think about your life, remembering that we only get one shot at life, so we better make the most of it.

Finally, I hope that by reading this book, you will consider having DBS. It's brain surgery, it's scary but it's not too scary, and it will absolutely change your life. I've gotten my life back thanks to DBS, and so I'm happy to sing its praises from the rooftops.

I look forward to you joining me on this journey as I recount my story. And please, if you ever have any questions or need guidance to help find the right information for you, write to me since I'd love to hear from you: nick@parkinsonswarrior.com.

Table of Contents

An Introduction to Deep Brain Stimulation

In this chapter, I'll provide an introduction to Deep Brain Stimulation from a layman's point of view. Remember I mentioned I'm not a doctor? So there won't be much in way of technical terms (though there are a few), and there will be no test after this to test your memory. I'm also not uninformed, so there will also be no "the thingy goes into the other thingy..." I'm somewhere in the middle with my knowledge, and you will be too once you're done with this book. You'll have enough information to have a good conversation with your doctor, in which the conversation is not lopsided, with them telling you everything and you just nodding your head. None of that. We Parkinson's Warriors believe that knowledge is power, and the more you know, the better. So let's get started.

What is Deep Brain Stimulation?

Let's get straight to the point, shall we? There will be much more on this throughout the book, but you should at least begin with some basic knowledge of what this surgery is, how it works, and how it can help you.

Deep Brain Stimulation is a neurosurgical procedure in which a neurosurgeon implants a neurostimulator in the body, which sends electrical impulses to the brain through implanted leads, targeting specific areas of the brain. In its simplest terms, DBS involves drilling a couple of small holes in your head, inserting very thin wires deep inside your brain, connecting the wires to a small battery and pulse maker, and then putting the battery / pulse maker into your chest.

Let's examine some of these terms:

Neurostimulator – Imagine this as a pacemaker for the brain. It is a small unit that goes implanted into the chest, controls the current to the brain, and holds the battery. This little guy sticks out very little from your chest, and is barely noticeable as a bump under your skin, especially for the newest models. It is usually implanted into the chest below the right clavicle. They usually put it on the right side in case you need a pacemaker for your heart one day, which must be implanted on the left side. Sometimes the neurostimulator is inserted into the abdomen area for safety reasons, but usually it goes into the chest. Also, if you have what's called a *bilateral surgery,* meaning you have the surgery done on both sides, you still only need one stimulator to run both sides of the system.

Lead – This is the tiny wire that is implanted into the deep part of your brain. The surgeon uses a probe to get the lead to the deep brain, and then takes the probe out, leaving the lead immovable so it can deliver charges to the brain. The lead contains the electrodes – the actual outlets of electricity, which can later be programmed by your neurologist or DBS nurse to maximize the benefit while reducing the side effects.

Extension – This wire connects the lead to the neurostimulator, under the skin. It sounds creepy, but the extension takes the lead connection and then routes it out of your head (under your scalp), down the right side of your neck (also under the skin), down through to your chest and into the stimulator in your chest.

Unilateral vs Bilateral DBS – In Unilateral DBS, the surgeon only installs the lead on one side of the brain, affecting only one side of the body. Since Parkinson's ends up affecting both sides of the body, most people end up with bilateral DBS, even if they only have one side done at a time.

Deep Brain Stimulation, or DBS, has been used to treat movement disorders like Parkinson's, essential tremor, and dystonia. DBS has been used for a couple of decades as an advanced treatment when traditional treatments no longer work as intended. In addition to being used for Parkinson's disease, DBS has been used to treat obsessive-compulsive disorder, epilepsy, chronic pain, and major depression.

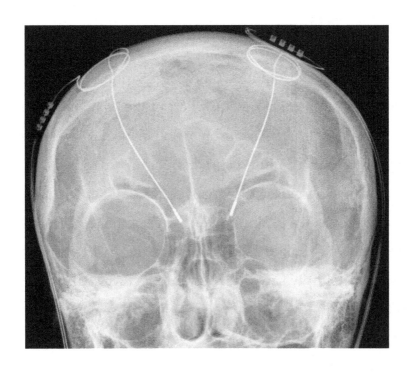

Figure 1: DBS-probes shown in X-ray of the skull. Hellerhoff / CC BY-SA
(https://creativecommons.org/licenses/by-sa/3.0)

DBS used to be a therapy of last resort for people with Parkinson's. It used to be the case that someone would consider DBS only after supplements, medications, exercise, and other non-invasive therapies had been exhausted. The problem is, if you wait until much later in the disease to get the surgery, then you would have spent most of your time with Parkinson's getting worse and unable to live your life to its fullest. These days, people are opting for this surgery at younger ages and at earlier stages of the disease.

My neurologist had been "hounding me" for a couple of years to have DBS. She would sing its praises and I kept saying "no, absolutely not, no." Part of that was me still not accepting that I had Parkinson's, and part of that was not wanting to undergo brain surgery. "How did I get here," I asked myself about my disease. At one point I even thought my neurologist was a shill for the drug companies, probably being paid to push unnecessary treatments. In the end, I was wrong. She simply saw what I didn't see. I was still grieving the fact that I had Parkinson's, and I was angry at the world and at this disease that had taken so much from me.

Finally, about three years of trying supplements and therapies that did little to mask the symptoms and caused incredibly painful side effects, my neurologist proposed DBS again. She said, "You only live once. You're still young. Don't you want to live your life to its fullest rather than just suffer for the next 30 or 40 years?" Wow. That hit me like a ton of bricks. A reminder that I only had one life to live, and I could actually decide how I would live it. DBS would give me back control of my life, even if just a bit. That did it for me. My medications weren't working as well as they used to because my disease had progressed so much, I was about 38 years old, and I still had a lot to live for!

If you decide to go for DBS surgery, the first step in the process is to check with your neurologist to see if you're a good candidate for DBS. I go into this evaluation process in a future chapter, but essentially you're a good candidate if you react well to Sinemet (carbidopa/levodopa), since DBS seems to act in the brain in a similar way to Sinemet. The bad news is that if you don't react well to Sinemet (it doesn't loosen you up and diminish your symptoms), then DBS is likely not available to you. The good news is, by having DBS surgery, in many cases you can almost eliminate the need for Sinemet.

Once your neurologist refers you for DBS, and you have undergone tests to ensure you are a good candidate, you will meet with the neurosurgeon to discuss the surgery. The neurosurgeon will inform you of the team's suggested target in your brain. The two main targets for DBS surgery in the brain for a person with Parkinson's, is either the globus pallidus internus (GPi) or the subthalamic nucleus (STN). These are simply the names of two places in the deep brain that doctors can choose to operate on. DBS surgery at the GPi target is known to improve motor function, while the STN target is used for people taking a lot of meds. Doctors typically avoid STN if you have a history of depression or impairment to your cognition, since STN surgery can increase depression, and it could also lower your cognitive abilities. With either type of DBS surgery, you're likely to experience an improvement of 30-60% in motor score evaluations (Wikipedia). Imagine 30% fewer motor problems, your tremor gone, being and to move your hands more and even walking better. This is just the low end of what's expected. You could have even better improvement, which is what the doctors saw in me after surgery.

Next, you get to choose your hardware (or often, it is chosen for you depending on the hospital). There are three main companies that develop and market DBS hardware. Medtronic is the oldest developer of hardware, and for many years they were the only option due to patents and their entrenchment in the neurological community. However, there are now competitors that offer alternatives for patients. Without the need to innovate for lack of competition, Medtronic's hardware is perhaps the most antiquated, though also likely the most reliable. The new kids on the block are Abbott and Boston Scientific. These manufacturers offer new options like rechargeable batteries (lasting up to 25 years instead of needing to replace it every few years), directional leads (which can be "steered" to avoid certain areas of the brain that produce side-effects), and state of the art remotes. Medtronic has begun to innovate again and they offer new and improved options. It is definitely a good time for anyone considering DBS.

Ultimately, it's worth choosing the hardware that your neurosurgery team works with and is most comfortable with. My team offered me any of the three options (some hospitals only work with one company), and after reviewing the features, I chose Boston Scientific's hardware. It simply made the most sense for me at the time. Despite this, I have no doubt that whatever option I would have chosen would have offered an improvement over not having any DBS device installed.

The doctors may recommend a unilateral DBS surgery, or a bilateral DBS surgery. If you have symptoms mostly on one side, or if you are a high risk patient, the neurosurgeon may suggest you only have DBS surgery to alleviate symptoms on one side of your body. If you are more seriously affected on your left side, they will implant the lead on the right side of your brain. Conversely, with worse symptoms on your right side, they will implant the lead on the left side of your brain. Don't ask me… that's just how the brain works!

If Parkinson's has affected motor function on both sides of your body, your neurosurgeon may suggest bilateral DBS, or DBS for both sides of your body. Sometimes it depends on your insurance – with yearly caps or insurance authorizations only allowing for one side at a time. I've met people who have had unilateral DBS surgery who are planning to operate on the other side in the near future. I was fortunate enough to have it done on both sides at once – bilateral DBS – so I had everything done at a single time.

One important note is that DBS only works for motor symptoms of Parkinson's. So the electricity delivered to your brain only affects the parts that affect movement, not cognition or psychological health. I used to take two Sinemets, four times a day, and I know people who used to take more than that, but after DBS I only take one Sinemet in the morning to help improve cognition, and one Sinemet at night to help treat REM Sleep Disorder (along with other meds). I can go, and sometimes have gone without taking a Sinemet for a couple of days and I have been fine relying solely on the DBS, albeit I do feel foggier those days. Deep Brain Stimulation also does not help with issues like balance, depression, or anxiety. It would seem that balance is a motor issue, but not in Parkinson's. So I can now run up and down the stairs, but I still need to grab hold of the railing!

The positive effects of DBS are incredible, yet DBS is not a cure. I and everyone else who has had DBS still have Parkinson's. However, now I have a tool that I can adjust and increase over time so that my symptoms will always be milder than without it. DBS is a tool in the Parkinson's toolkit, and should be seen as such. Many people with DBS systems they had implanted years ago are back to taking meds, just at a lower amount. DBS does not stop Parkinson's progression. We still need a cure!

The exact way that DBS works is not completely understood (there's still a lot we don't know about the brain), but generally, doctors think that by inserting the lead into the area of the brain that controls movement, DBS either a) blocks the distorted signal that causes Parkinson's symptoms, b) activates neurons that help regulate Parkinson's symptoms, c) sends its own signal to disrupt Parkinson's signals, or d) blocks Parkinson's signal throughout the brain. In layman's terms, DBS essentially sends an electrical current to the area of the brain that controls movement in order to trick your brain into thinking the signal is correct and not distorted by the disease.

Although DBS involves putting wires deep inside the brain, DBS is considered to be a non-destructive procedure.

In fact, doctors have said that DBS can be undone once there is a cure for Parkinson's. Although it's not a perfect solution to Parkinson's, it's better than what doctors used to do in the past. In the past, therapies were much more invasive and destructive. One such therapy involved **pallidotomy**, a neurological surgery in which doctors implanted wires in the brain, but instead of connecting the wires to a stimulator like they do today, they would raise the temperature to 176 degrees Fahrenheit (80 degrees Celsius) and actually burn a part of the patient's brain! Another procedure called a **thalamotomy**, destroyed part of the thalamus (deep in the brain) by cutting or burning it as a way to treat tremors.

These procedures seem barbaric by today's standards, but they helped patients with levodopa-induced dyskinesia and tremors, respectively. These procedures of course had downsides, side effects, and unintended consequences. But it was the best they had at the time, and it's through that work with those patients that lead doctors to the DBS we know today. I'm sure that even today's DBS will seem barbaric to future generations. Hopefully, we'll have a cure and won't need DBS at all one day, but for now it's the best we've got.

Back to the modern DBS. Remember that DBS has three main hardware components: The neurostimulator, the lead, and the extension. All three components are implanted surgically inside the body – the system becomes a completely internal solution. The surgery can happen under general anesthesia (known as "asleep DBS"), or half awake and half asleep (traditional DBS), in which the doctor first inserts the leads and tests them with the patient awake, and then the patient is put under for the remainder of the surgery. To get the lead into the brain, the surgeon makes the head immovable with the use of a stereotactic frame (which almost looks like your head is inside a couple of tennis rackets), drills a hole (if unilateral) or two holes (if bilateral) into the skull, and uses a probe to insert the lead directly into the target.

During traditional DBS surgery, the patient is awake for the first part (drilling, probing, placing the lead), since the neurosurgeon depends on the patient's feedback for the correct placement of the lead. When the surgeon is confident they have the lead properly placed, they put the patient to sleep under general anesthesia to install the neurostimulator and connect the extension under the skin from the head to the chest.

It used to be the case that all patients were awake during the first half of the surgery – this meant you were awake when they drill one or two 14mm holes into your skull and insert the leads into their proper locations. This part is done awake so that doctors can know if the lead is properly placed and is not causing side effects. Once the lead is properly placed, the patient goes under general anesthesia for the remainder of the surgery, when the surgeon closes the wound on the skull, routes the wire or wires from the leads across the head and down the neck into the chest and into the neurostimulator. Depending on your insurance and in what country you are having surgery, the neurostimulator may be implanted on the same day or on a different day.

So what happens during the "awake" part of surgery? First, the patient is prepared for surgery. The most important part of this preparation is the placement of the stereotactic frame on the patient's head. This is a large metal frame screwed into the patient's skull (local anesthesia is placed on the contact points). This is so the head is immobile during surgery. This is important, because often, a brain scan has been taken before the surgery or one is taken direct before the surgery. This is so the surgeon knows where and at what angle to insert the leads into the brain. Since the head is not able to move, the prior scan and current scan will match up with the patient's current head position. By this point, the surgeon has studied your brain scan and has a plan to reach the intended deep brain target with their probe while avoiding major arteries and veins in the brain.

When everything is set, and the operating room is ready, a drill is set and the surgeon initiates the drilling through the skull in order to expose the brain. The drill creates a hole of approximately 14mm in diameter (smaller than a US dime). No need to be afraid – although patients claim the "noise" in their head during the drilling is likely the loudest sound they have ever heard, a) the area is anesthetized so there is no pain, b) the drill knows when to stop automatically, so it's impossible for it to go further than the skill, and c) the process only lasts about 10 seconds on each side, so by the time you realize what's happening, it's over.

Now the doctors match up the brain scans and line up the stereotactic frame to precisely insert the lead into the target. When they're ready, they attach a probe with the lead to the frame and guide the probe through the brain and into the target area of the brain. The brain has no pain receptors, so this is painless and the patient is often unaware that it's even happening. The patient will be asked to make a fist, touch fingers together, speak, and perform other tasks to ensure the proper placement of the lead. When the surgeons are satisfied that the lead is correctly placed, they give the patient general anesthesia and perform the remainder of the surgery with the patient asleep.

Here is what happens during the "asleep" part of surgery. The wound on the patient's head is closed – the hole in the skull is secured with a burr hole cover, and the hole in the flesh is closed with stitches. This burr hole cover is no regular patch job. This cover holds the leads in place and protects the skull after surgery. You can roll around in bed, snuggle your head into your pillow, and rest assured that your DBS leads are safeguarded – nothing on your skull is moving after this surgery.

The surgeon then takes the extensions connected to the leads, and routes them around the head (under the scalp), down the neck (under the skin), and typically (though not always) into the chest area below the right clavicle, where the neurostimulator is installed and the wires are connected. For some patients, the neurostimulator is installed in the abdomen, either for safety or for aesthetic reasons. Then all of the wounds get stitched shut. In my surgery, the head wounds had metal stitches, and the chest wound had dissolving stitches that came out on their own after about a week. The patient is bandaged and then sent to the recovery room until they wake up from the anesthesia. The surgery is now over!

New technology has allowed for the entire surgery to be conducted "asleep". This technology has been perfected, and I expect most, if not all DBS surgeries will one day be done entirely under general anesthesia. If DBS is done fully asleep, like mine was, the patient is put under right from the beginning. The way this works is that the surgeon matches a previous scan of the brain with one taken directly before surgery while the patient wears the frame. The surgeon matches up the scans and can hit the correct target without feedback from the patient during surgery. When I had my DBS surgery, it was done fully asleep and the post-surgery scans showed that the leads were perfectly placed. I never had to go through the awake part of the surgery, and yet they knew the surgery would work, which it did. Trust me, if you're given an option, choose the fully asleep version – you won't remember a thing!

Once the surgery is over and the patient has had a chance to recover (usually 10-21 days), they visit with their movement disorder specialist (or more commonly, a DBS nurse) to provide an initial programming of the neurostimulation device in their chest. During this session, which can take hours, the leads and their electrodes are all tested to ensure they are working properly and to check how much voltage the patient can endure before side effects occur. The patient leaves the appointment with an initial program setting that generally works well for most patients. They are also given a remote control, which allows for restricted increasing or decreasing of the stimulation amount. Over the next several months, the programmer fine tunes the stimulation for more personalized settings. These are done as outpatient procedures at the DBS clinic. Once the stimulation is turned on it runs continuously and is never turned off. The only time the neurostimulator is turned off (either by a doctor or with your remote control) is to change the battery or to make the stimulator safe for MRIs and other brain and body scans. Otherwise, the stimulation runs 24/7 without interruption.

While I had no complications and minimal side effects due to the surgery – some minor short term memory problems that seem to have mostly resolved themselves, though not completely – not everyone has the same result. Remember, everyone with Parkinson's has different symptoms at different severities. This means that a surgery like DBS is going to have varying results for people at different stages of the disease. For me, it went great. I lowered my dose of Sinemet significantly, I could type again, roll around in bed, walk, and even have more strength and energy. My neurologist told me my improvement was 350% compared to being unmedicated. Everyone told me this result was remarkable and not the most common outcome.

The most common outcome is to be as well with DBS as you are on your Sinemet. Depending on the severity of the illness, this can be a 200% improvement, or even a 100% improvement. Regardless of your prior symptoms, most (95%+) patients have a positive result and experience improvements. A small percentage of people experience little to no improvement, and a small number of people experience some complications due to the surgery. Remember, this is brain surgery. Most top surgeons have done this surgery hundreds or thousands of times, but things can still happen. Some complications include infections (on the surgery wounds), aneurisms, and strokes. Again, these are extremely rare, but they do happen. Ask your neurosurgeon what their complication rate is. In Amsterdam, where I had my surgery, the doctor pointed to one complication (an infection) in the hospital's thousands of surgeries. That's a complication rate of 0.04% assuming 2,500 surgeries (my estimate). That's a great number!

If your main concern is complications due to brain surgery, know that DBS is generally a safe and relatively routine surgery, and your potential gain is much greater than your potential risk. Just to give you an idea of how I feel about this, If I had to have brain surgery every year to regain the benefits of DBS, I would do it without blinking. Luckily you only need to have surgery once (or twice if you have one side done at a time)!

To lower the most common complication (an infection at the surgery location) some doctors recommend their patients keep their bedding clean and take showers daily. This helps lower post-surgery infections significantly. I spoke to many doctors around Europe before deciding to have surgery in my then-current city of Amsterdam, and the doctors all told me that complications involving infection all happened at home for the patient, due perhaps to a failure to change the bed sheets daily or not shampooing daily.

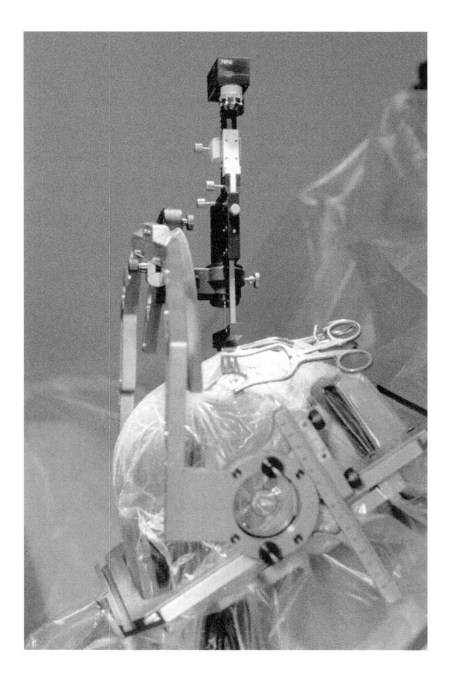

Figure 2: Insertion of an electrode during deep brain stimulation for Parkinson's disease. Wikimedia Commons

Alternatives to Deep Brain Stimulation

While DBS is now a go-to therapy for intermediate to advanced cases of Parkinson's, and while DBS is safer than it has ever been in the past, it is understandable that people with Parkinson's will still seek out less invasive alternatives to this therapy. I know that for myself, it took a few years and several mentions of DBS from my movement disorder specialist before I began to warm to the idea. These alternate therapies include advanced medications (that are stronger and with more side effects than Sinemet), and devices like the Duopa pump, which is also an invasive therapy, but that is seen as a better option among many patients who would still like to avoid brain surgery for a while longer.

Keep in mind that Parkinson's is still incurable, and that no medication or medical device can stop the disease's progression. After a period of time, which is different for each person, therapies stop being effective. The best we can do right now is delay the onset of the severest symptoms. After Sinemet has reached its limits of efficacy due to the progression of the disease, people with Parkinson's turn to stronger medications with more severe side effects. These include a wide array of medication classes, but one of the most common type is **dopamine agonists** like apomorphine, bromocriptine, rotigotine, pramipexole, ropinirole (all generic names).

Another DBS alternative is a **Duopa pump**. This is a device that is installed surgically into the abdomen and through to the stomach, and that delivers a continuous stream of a carbidopa/levodopa gel suspension directly into the intestine. This therapy helps prevent on/off periods and helps reduce dyskinesia. However, the side effects and complications are not trivial (could include infection at the surgical site), and should be weighed when considering Duopa as an advanced therapy.

Therapies for Parkinson's are always improving, as there is a lot of research (and a lot of research grants) associated with finding better treatments, and ultimately a cure. There is promising work being done with stem cells, with existing medications being used off label, and with genetic therapies using CRISPR technology. Search the web for these terms to find the latest research in these areas.

Ultimately, it's up to each person to decide, and then to work with their doctors to determine the best course of action. Obviously, I have had great results with Deep Brain Stimulation, thus my preference of it as a treatment and this book about the topic. However, I can easily see books being written by ecstatic people with Parkinson's writing all about the Duopa pump, or about any other therapy that has worked for them.

Listen, brain surgery can be scary, and searching for alternative therapies is normal. I had never been operated on, and this made things extra terrifying. It's normal to be hesitant, and to weigh all of the options. Even though I knew logically that things would be alright, emotionally I kept thinking about the potential complications of having surgery. People die during surgery, even routine surgery. Even though this is not normal, I couldn't help but to think I might die. But at the moment I decided on DBS, I thought "am I willing to die to get the relief that DBS could bring?" Could I face the worst case scenario to have a better life? In that moment, I said yes. From that moment, DBS surgery day couldn't come fast enough!

The Story So Far

Much has already been written about my Parkinson's story, primarily in my book, *Parkinson's Warrior: Fighting Back and Taking Control*. However, I'll briefly be retelling my story here with some updates.

In 2009, at the prime of my academic teaching career and at the pinnacle of health (I played tennis several times per week), I began having stiffness in my left arm. I was 31 years old, and I could not imagine that the issue was anything more than a strained muscle or a pinched nerve. I went to visit the doctor at my HMO hospital, and he examined my issue. My left arm appeared to be very tight and frozen in one position, appearing as it does when you have your arm in a sling. The doctor suggested it was probably from too much tennis, so he sent me home with some ibuprofen and instructions to rest my arm for a few weeks.

After a few weeks, the tightness didn't go away, and in fact now my left leg began to tighten so that I couldn't rest it, even in bed. Going back to the doctor, he was still at a loss so he referred me to another doctor of general medicine for a second opinion. This second doctor also thought it was all inflammation from playing tennis. He gave me a prescription to take three 800mg ibuprofen capsules per day for up to six months. I went home again with the hopes that this health issue would simply go away.

As you can probably guess, after six months my condition didn't improve but instead worsened, with new symptoms including loss of smell, a very low voice (which was embarrassingly brought to my attention by a student – "speak up, Mr. Pernisco!"), and my left leg was now dragging. I wrote to my doctors, and after much kicking and screaming (again, this was an HMO, organizations that I believe are hellbent on reducing costs at the expense of patient health), I was finally referred to a neurologist.

The neurologist did a neurological examination, and at the end of the session he told me that what I likely had was Parkinson's disease. This was his working diagnosis, but not one that even he took seriously. "How can someone so young have this disease?" This was the question on his lips as he sent me to see a second neurologist to confirm. I had no idea what Parkinson's was. My first question was, "will I die?," which was a stupid question and poorly phrased at that. Instead of listening to his answer, I decided to do some research at home. I've learned a lot about the subject since then.

Parkinson's disease, as it turns out, does not kill you. In its most common form, it makes your muscles rigid, makes your hands tremor, and makes you move very slowly. None of these symptoms affect your organs like some more aggressive neurological diseases do, so it won't affect breathing or heart beating. Of course, people with Parkinson's do die, of old age or of complications like taking a bad fall or aspirating food or liquids into their lungs, leading to choking or pneumonia.

What Parkinson's disease would do to me, however, was shorten the amount of time I could work, and even limit the type of things I could do. In the short term, symptoms

27

could be treated with medications, but in the long term, I would likely end up in a wheel chair and need round-the-clock assistance. There is currently no cure for Parkinson's, but I'm hopeful that something will come along before that late stage arrives.

To get back to the story, I went to see the second neurologist, and he concurred with the first neurologist's diagnosis: it's likely Parkinson's. He said "likely" because, again, "how can someone so young have this disease?" I wanted to know for sure, so the neurologist referred me to the only movement disorder specialist that this HMO had on staff in the area. This guy had already retired, and worked only a couple of days a week to help in cases like mine.

After visiting with him a couple of times, and after a wide array of blood tests to rule out any other possible diseases, he finally told me I would get a PET scan to confirm the diagnosis. The procedure involved me taking radioactive tracer pills and being put inside of a giant MRI-type machine. My test was delayed by several months because the only lab in California that made the pills was behind in making them. But after many months, I had the scan done, and the results were in (by e-mail): "Your scan is consistent with that of a person with Parkinson's, so we will proceed as such."

That was in December of 2011. I finally knew: I had Parkinson's.

Instantly, my world turned upside down. I was overcome by a sense of fear about what the future would hold for me. Would I be able to do all of the things I wanted to do? Would I be able to travel? For how long would I be able to work? And most scary of all, would I be a burden for my family in the years to come? I wasn't going to die from Parkinson's, but how would I live with it?

Within a week or so of that pivotal appointment, I went out and bought myself my dream car: a sports car with a big, powerful engine.

I would begin to live my life this way; "I don't know what the future holds, so I want to experience this now." I also had a tough time accepting that the diagnosis was something I couldn't just fix by ignoring it. I still had a lot to prove to the world, and to myself, and I had little time to do it before the wheelchair came calling.

I began walking and jogging for exercise daily, I fell into the trap of taking supplements as a way to forestall taking medications with harsh side effects, and I would do foolish things like go back to school just to gain approval from others that I was still "worthy of being paid attention to." This lead to some serious anxiety and depression early on, but things would get worse before they got better.

I was 33 at the time, so at a time when most are just getting started with their lives, I was forced to begin winding down and to start thinking of what the end might look like. I was assured that symptoms would only get worse over time, and that medication would only mask symptoms. According to doctors, there was only a decline in my future.

I was at my best in terms of my skills and knowledge – I was teaching college classes, I had already run two successful companies, and as my brother later told me, I had nothing to prove to anyone. But I felt I needed to continue proving myself, to myself and to others. This was a part of living with denial. I thought it was as simple as mind over matter. Unfortunately, things wouldn't be so easy.

Before Diagnosis

Let's backup for just a moment. I want to tell you about my life before Parkinson's, which may put some things into perspective, including why I would eventually opt for brain surgery. In fact, let's start at birth.

The story goes that, when I was born, I was not breathing and I did not have a heartbeat. I was given shocks to revive me, and after a moment, I was alive and crying like every other newborn that day. One of the nurses told my mom, I may have died if not for being born in such an advanced hospital like UCLA medical center. That was in 1978. My older siblings were born at a different hospital, so my parents always had a "he was born and lived for a reason" kind of mentality about me and my life.

I grew up in a middle-class home, the son of immigrants from Argentina. As the youngest (and perhaps the most fragile) of three, I was used to getting a lot of attention. This no doubt fueled my life-long desire for getting other people's approval, though this would eventually change when I learned the rules of happiness (rule 1: happiness comes from within).

I was very tech savvy from a young age. Some of my first memories include having had an Atari 2600 in the house (around 1980), and it was when I was around six years old when my parents bought a Coleco Adam computer for the family. I would go on to have every major gaming system of the 80s, and several more computers, including my old favorite, the Commodore 64, which I first learned to program on at around 10 years of age. Those around me and my family thought I was a whiz with computers, though by today's standards, I was just a "power user." A lot of expectations were set on the youngest of the family... go to school, get good grades, and do something with computers.

My family had a business in Argentina when I was a teenager, and I would travel with them seasonally during the American fall and winter, which was the Argentine spring and summer. Because of this, I was homeschooled until I was about sixteen years old. Calling my schooling "homeschooling" is generous, actually. It was more like I was self-taught. I was given free rein to design my own curriculum, and it included computers and the arts. Mostly, this meant I programmed a lot and I watched a lot of movies.

When I arrived to college (by way of testing out of high school instead of graduating), I majored in computer science. My first programming class was in a language called COBOL, made for mainframes and of special note at the time due to its role in the infamous Y2K bug. I loved the class, and when I researched the college catalog for which classes to take next, I noticed a lot of courses required high levels of math. Unfortunately, math was not a subject I had dedicated much time to in my self-made curriculum, and I realized after taking some more advanced classes that it would be too much of an investment in time and energy to get to the levels of math I needed, so I changed my major.

Taking courses in different subjects to see what I would like best, I eventually settled on radio-TV-film production. Instead of a computer whiz, I would be a media producer, perhaps even a film director. I eventually went on to earn my master's degree in screenwriting, and this would soon lead me to start teaching. I taught the courses that I had aced as a student, at both my community college and at my undergraduate university. I fell in love with teaching, helping students learn what I had learned years before and making sure they were ready for anything in the media business. Around this time I also started a film production company, making documentaries and educational videos that would go on to be distributed internationally. Film was a big part of my life after graduation and until I was diagnosed with Parkinson's. I wrote, produced, and directed about 12 films in the span of about six or seven years. I also produced music, and co-hosted a podcast with a friend. I absolutely loved the work, but it all came to a screeching halt when I was diagnosed.

With Parkinson's, I could no longer teach in person – I was weak and my voice was too soft. I could also no longer work collaboratively with other creators, since my motor symptoms were severe early on – slowness, rigidity, tremor – all ended my filmmaking and music producing careers. This was a great loss for me, and years later I would try to fill the void by going back to school, and constantly trying to prove myself.

When I look back at the course of my life, I now feel that Parkinson's was there with me all along, perhaps even from that initial zap when I was born. I never did well at sports, always getting tired quickly and never able to keep up. Even tennis, which I had become very good at playing, I was worn out after just one set (about an hour) and always noticed my legs felt heavy when I played. My father worked as a house painter, and whenever I would go work on a job he needed help with, I would get tired very quickly, needing lots of coffee and a lot of breaks just to make it through the day. Back then it could all be chalked up to being a lazy 20-something, which I suppose made sense to some degree, though I wouldn't consider myself lazy when doing the things I loved. Even though I loved video games, I was never the best at them, my fingers were too slow or my movements were too sloppy to perform the "magic move" that everyone else could do to win the game. Even playing pinball, perhaps my favorite pastime of all time, I noticed I was slower and less coordinated than many of the kids I played with. Doing just ten pushups would make my arms shake, and lifting weights would always leave me sleeping in late the next day.

Looking back now, I was probably compensating for Parkinson's even in childhood. I loved computers and games, but not being good at games, I decided it would be more fun to do something slow and thoughtful like programming. When I worked to produce music or shoot a film I had written, it would always be on my terms, on my schedule, on my timeline. This meant smaller productions with fewer people involved. Teaching worked out well for me because, not only was I an expert on everything media, but it was something that only took about 10-15 hours per week to do, leaving lots of time for (as I see it now) rest.

I came to the realization that I had been subconsciously self-selecting activities my entire life. I was choosing activities and directions in my life that avoided highly stressful situations, increasingly debilitating fatigue, and symptoms that felt like normal aging at the time but I now realize could have been Parkinson's all along. This realization gave way to a mind-blowing question that would take me years to answer: did I *ever* know what it felt like to be unimpaired?

Shortly after my diagnosis, I thought back in my life to what could have caused Parkinson's, not that it mattered or would change anything, but we always try to justify or understand why something happened. Everything was fair game, since Parkinson's symptoms could take years or decades to appear. According to the research, having had repeated concussions is a precursor of Parkinson's. Thinking back, had I ever had any concussions in my life? I was hit on my head with a set of keys once as a kid, and another time I was hit on the head with a rock, also as a kid – both times my head had bled but I had none of the effects of a concussion: headache, ringing in the ears, nausea, vomiting, fatigue or drowsiness, or blurry vision (Mayo Clinic). Other research has shown that a sudden blow to the head could be a precursor of Parkinson's. Had I ever had sudden blows to the head? I hit the back of my head twice before, once ice skating and once on a slippery concrete floor, and one time I was hit in the face with a soccer ball going full speed, but I don't remember having any symptoms of concussion or any long lasting effects. More research has shown that exposure to heavy metals like mercury or lead, exposure to pesticides, and a family history of Parkinson's are all precursors and could indicate Parkinson's in someone's future. I had not experienced any of those, and no one in my family had Parkinson's.

Ultimately, I gave up trying to figure out why I had Parkinson's. Even if I discovered the reason, I already had the disease and it is incurable. It would be best for my mental health to not dwell on things in the past I couldn't change.

The Decline

Michael J. Fox famously told his doctors to just give him the medications in response to his Parkinson's diagnosis. Of course, as we all know, the medications only mask the symptoms of Parkinson's, but don't do anything to modify the progression of the disease. For years, people with Parkinson's take the medications meant to mask their symptoms, which help them live better lives than they could otherwise without them. But medications come with their own problems. The side effects of some Parkinson's drugs are almost as bad as the symptoms they treat, giving patients a difficult choice when deciding on the trade-offs they're willing to live with for some sort of alleviation of symptoms.

Like most people with Parkinson's, I began my life with Parkinson's by taking the standard concoction of carbidopa/levodopa, also known as Sinemet. The so-called "gold standard" medication for Parkinson's is a Sinemet tablet that is 25mg of carbidopa and 100mg of levodopa. Human anatomy has evolved to protect the brain from foreign invasion, creating a blood-brain barrier to prevent foreign debris from endangering the brain. The carbidopa is used to help break through this blood-brain barrier and get the levodopa through so it could do its work. The job of levodopa is to give the brain the dopamine it needs to allow movement. Ideally, if Sinemet works correctly, then a person with Parkinson's would have no physical symptoms.

This scenario sounds great. I could just take Sinemet and the symptoms would go away? As Michael J. Fox said, just give me the meds! But not so fast. Sinemet can have really awful side effects like nausea and vomiting in the early years, and over time people develop excessive movements called dyskinesias. The best way to describe what dyskinesias *look* like is that you're itching all over and constantly wiggling in your clothing. What dyskinesias *feel* like is the sense of being unbalanced on a small boat, constantly trying to balance yourself using your entire body.

Dyskinesias are the reason that many people with Parkinson's try to avoid taking Sinemet early on in their disease. They know that as soon as they begin taking Sinemet, they have begun down the path of taking an ever-increasing amount of carbidopa/levodopa. Over time, the medication loses its efficacy and can no longer keep you in an ON state because your disease has progressed so much. This means you eventually has fewer ON periods, and these ON times last a lot less, meaning you need to take larger and more frequent doses of Sinemet as time goes on. More Sinemet means more side effects, including disabling dyskinesias. This cycle perpetuates itself until you're taking, as an example, twenty pills of Sinemet each day in four or five different doses, just for a little bit of relief. It's no wonder then that a lot of people would prefer to delay taking Sinemet for as long as possible.

I still remember the first time I took Sinemet. I had only been taking natural supplements until that time, trying to avoid getting on that Sinemet cycle so that I could save the Sinemet for when the disease worsened. The main issue with that was that I was still young when I was diagnosed, and I didn't want to spend my 30s being ravaged by the disease. Seeing that the cocktail of natural supplements wasn't working for me (at one point I was taking over 20

41

pills of supplements each day), my naturopathic doctor gave me a pill of Sinemet to try. That moment blew my mind, as it was the first time I could remember having some relief from this disease. I asked my naturopath for a prescription, and she agreed as long as I kept up with some of the supplements that she recommended that had antioxidant qualities.

Taking Sinemet was a life-changing experience. After a couple of years of increasingly severe symptoms, suddenly my body felt freer than it had in years. I could move my arms and hands, and I could even walk normally. It was a dream come true, or better yet, the end of the nightmare. My only regret at the time was to not have started taking Sinemet earlier. The medication wasn't a cure, but it changed my life for the better.

I was riding high on the Sinemet wave for about two years when I started noticing changes in my movement again. Suddenly, the rigidity and slow movements were coming back. At first my doctors suggested increasing the dose and frequency of Sinemet. We tried 1.5 tablets, every four hours, then 2 tablets every six hours. The fine tuning process was difficult, as I was never receiving the optimal dose when our various experiments failed.

The on-off periods became so severe that eventually, my doctors recommended a controlled release version of Sinemet in addition to the regular dosage. The instant release would take effect within an hour of dosing, and the controlled release would take effect more slowly and reduce the off period between doses. This worked to some degree, but then the dyskinesias came.

No one is able to escape the conclusion that Sinemet will eventually lead to dyskinesias. In my case, the dyskinesias were so severe that having them was considered a disability. My left leg jumped every time I made a stride, very similar to the characters in the Monty Python skit *The Ministry of Silly Walks*. I was unable to keep my leg from jumping up like this, and walking became unbearable. It took so much energy to move myself during this time. Walking a block would take half an hour, and by the time I was done I was extremely tired and I had sweated as if I had just had a huge workout.

If the dyskinesias weren't enough, my on-off periods ruled my life. I would have to wait and be ready for my medications to take effect in order to shave, brush my teeth, shower, type on the computer, and do most of my everyday chores. And when I started doing anything, I only had a specific amount of time before the medication wore out and I was off again. I even began having failed doses – when I would take my medications as normal, but the dosing would never take effect. So in essence, I would take my Sinemet in the morning, wait until it took effect to shave and shower, but the on-period would never arrive. Usually, this only happened in the morning, but it also occasionally happened in the afternoon, and at night all of the Sinemet I had taken for the day would take effect all at once, causing dyskinesia.

Needless to say, my quality of life at this point was at the worst it had been in a long time. I couldn't function as an independent adult. A few times, my wife Rosaline would need to come home from work because I wasn't doing well and couldn't do anything for myself. Although I had the Parkinson's Warrior mentality and attitude to get me though, I couldn't help but get depressed at times. I was usually able to break myself out of the funk by doing some stretches or listening to music, but I wondered if this was how bad it was going to be from now on.

My doctors wanted to continue trying other medications, especially the so-called dopamine agonists. This is where I would have to make trade-offs. I could try one of these agonists, and possibly have smoother movement, fewer dyskinesias, and perhaps less off-time, but in exchange I would sacrifice some cognition and memory, and I would need to make peace with one of the worst side effects I had ever heard of. See, many people would develop compulsive and addictive behaviors while taking dopamine agonists. People who took this type of medication could become addicted to sex, with many seeking the services of prostitutes. Others would become compulsive gamblers, gambling away the family savings or home. Others would become addicted to drugs or alcohol. I had heard stories in support groups about how these medications had ruined people's lives.

I refused to take them.

The only other option was Deep Brain Stimulation surgery. At first when my doctors recommended DBS as a solution to my dyskinesias and on/off period problems, I refused to even consider it. It's scary. There is a chance, albeit a very small chance, that I could have a complication – an aneurism, a stroke, an infection, even death. I knew that the chances of having one of these complications were unlikely. But it still scared me. I decided, the only way I would be fine with having DBS surgery was if I accepted that I could have a complication, and that I could die during surgery. Was I at the point that I would risk my life for relief from this disease? That was the question I had to answer to know I was ready for surgery. I don't need to tell you how I answered it.

The Decision

By late 2017, my physical Parkinson's symptoms had become unbearable. I was stiff all of the time, I could barely walk, and I thought things were going to end any day. My emotions were all over the place. I was still recovering from the huge disappointment of leaving graduate school, and to say I was incredibly anxious and depressed would be putting it mildly. The optimistic spirit in me was almost fully depleted, and if not for the small bit of internal fight I had left, I probably would have lost all hope.

The main reason things were so bad was because my medications were not working like before, and I had been having severe on/off periods, and they were happening several times daily. I would wake up in the morning, barely able to move – it would sometimes take me 30 minutes to just muster the energy to roll out of bed. I would take my morning pills, and within an hour I was ON. This meant the meds had taken effect, but the ON was so on that I had side effects called dyskinesias. Even with just a single 25/100 tablet of Sinemet, I could move freely, but my body would also move on its own uncontrollably. Then, within an hour or so, the medications would wear off and I would once again be frozen, unable to move my body or use my hands. Then I would take medications again to have another on period. Imagine going through this every day, trying to plan your life around on periods that felt *too* on and sudden off periods that left me immobilized. I had to schedule out my work for on periods, schedule my shaves and showers for on periods, and generally live according to these on periods, which would ultimately come and go as they pleased.

I cried to my neurologist about this, and she suggested I have deep brain stimulation surgery. This surgery would replace my main medications and would make my on and off periods very smooth and barely noticeable. My doctor discussed the procedure with me. It all sounded so scary, but what was my alternative? I had grown tired of the way Parkinson's had been treating me, and I was ready to give up. At the time, I was still developing my Warrior mentality, but I hadn't prepared myself for such an intense fight so early in my life. I saw DBS as a way to increase the positive experiences in my life; to have more meaningful interactions with Rosaline, with friends, and with family. So I put myself on the list for the surgery at Seattle's Swedish hospital, which had the finest Parkinson's center in the Northwest.

But soon, life would intervene.

Rosaline was doing great at work, so great in fact that her upper management offered her a role in Amsterdam, to start ASAP. Wait a second, what about my surgery? I would need to wait at least a couple of months to be evaluated, then surgery, then recovery and fine tuning, which could take another six months. I was at the end of my rope, and was conflicted, but I saw this as too good an opportunity for Rosaline, and if I were to die, at least I would have checked off "live abroad" from my bucket list. Rosaline promised me we would get set up with doctors and surgeons right away, and of course I was being overly-dramatic about dying, so I agreed.

Amsterdam was a beautiful city, but I spent my first year there not doing much because I didn't have the energy, stamina, or strength. There was so much to see and do, but most activities involved walking. I learned the public transport system well, and this helped me join monthly book clubs and writing groups, but I could barely make it to the beautiful parks or stroll through the exquisite museums and galleries all around town. Immersing myself in the culture was impossible, since whenever I made it to a bar or café, I would speak too softly to be heard or I would mumble my words. It was frustrating for me and for those I was trying to connect with. After a while I gave up on trying to socialize and went to the cafés on my own and kept to myself.

Since the Netherlands has what amounts to a socialized medicine system, more like the United States Affordable Care Act (Obamacare) than Medicare for All (more of a British- or Canadian-type single payer system), I had private insurance that cost around $220 per month but covered everything from medications, to dental and vision, to specialists and operations. The Dutch medical system negotiated drug prices and availability for the entire country, and also subsidized certain common medications. Because of this, one of my main medications was not available in The Netherlands – they must have decided it was too pricey, so no medication – and this made me worse off. In the US I'd take nadolol, a beta blocker for high blood pressure, but it also had properties that helped control tremors. This was replaced in the Netherlands with propranolol, which worked well for the high blood pressure, but which brought back my tremor. Another of my medications, Sinemet CR (controlled release), which I took in addition to the regular Sinemet, was not very common in the Netherlands, and I had to fight just to get a prescription filled. Eventually, my neurologist helped figure out a medication regiment without the CR, but it didn't work as well.

Most days, I felt like staying in bed, but I would eventually get up, do some teaching work, and make it a point to exercise. At the time I was also learning to become a Parkinson's Warrior, and I was writing my first book. I would often walk daily for about 30 minutes, usually during my off times so that I wouldn't have that terrible leg dyskinesia, and I would end up at one of the nearby movie theaters and spend the afternoon watching movies (I had an unlimited pass). After watching a couple of the latest films (in English with Dutch subtitles), I would return home and lay down for a nap until Rosaline came home. This was my routine for nearly a year.

Although Rosaline's work was going well, our home life wasn't. I was such a burden on her that she would often come home early from work because there were days I couldn't feed myself or go up and down the stairs in our apartment. I wasn't doing well, and everyone around me knew it. I always looked tired, I always *felt* tired, and I was dragging both of us down.

Luckily the doctors at the Amsterdam Medical Centre realized that I was in really bad shape, and after I was told that it would be an 18-month wait for surgery, they called me in after only nine months.

The Surgery

In December 2018, I finally had Deep Brain Stimulation surgery to alleviate the symptoms of Parkinson's disease. It was quite the experience, and my family and friends were in awe, not only of how I handled it but also of the transformation afterwards. This was a life-changing experience, and so I would like to take you through a timeline of my surgery so you can get an idea of the process. The surgery happened at the **Amsterdam Medical Centre**, in the Netherlands, and I honestly couldn't be happier with their support or care.

September – three months before the surgery

The hospital called me and wanted me to undergo an evaluation. This is a weeding-out process to ensure the surgery turns out well and has a benefit for the patient. Entering the hospital was a bit intimidating at first. It was larger than any hospital I had been to in the United States, but my mind was put at ease when I arrived in the atrium and saw they had a Starbucks. The wonderfully-positive effect that a little comfort from home can have on someone is amazing. I would become known by the employees at that particular Starbucks, since they'd see me before any of my appointments over the next six months.

Getting operated on in the neurosurgery center required approval from a panel of doctors. They would examine me, ask me questions, run lots of tests, and generally see if I was a good fit. After these tests, I would receive the decision of whether I would be having the surgery or not. These doctors had some questions they needed answered before I could go through with the surgery – was I healthy enough for surgery (could my heart handle such a stressful event, did I have sufficient cognition to give up a bit of it, since DBS surgery could negatively affect memory), and would I benefit from the surgery (was I reactive to Sinemet, was I in the *sweet spot* – not too early in the disease and not in the final stages)?

A nurse called me one day and gave me the date to come in. This is how it worked in the Netherlands; they give you the date and time of your appointment and you nod your head and oblige. This grouped scheduling was especially important for this qualifying testing, since I would be seeing several doctors in different disciplines over a two-day period, which included an overnight stay in the DBS center. Over the course of two days, I had the following tests.

A **cognitive** test, testing my logic, memory, and spatial abilities through a variety of written tests. Some people who go through DBS lose some cognitive ability afterwards, so they want to make sure I could stand to lose a few IQ points during the process. Remember, DBS is brain surgery, and the inserted leads pass through the brain itself, causing damage. I have always prided myself on my high IQ, and I saw the cognitive tests as a competition in which I was trying to beat the test makers. The logic and spatial tests were easy and I easily scored in the 98th percentile of test takers for each. The memory test was the final test, and probably the most fun. I was given 15 objects to remember, then distracted with some other tasks, then returned to the test and I had to remember the words that had been dictated. The nurses administering the memory test were flabbergasted that I could recall all 15 objects in order, after 20 minutes of doing something else. I scored in the 99th percentile of people taking that particular test. So the answer here was "yes," I could stand to lose a bit of cognition after the surgery.

Next up was a **psychiatric** test, testing to see if I was sane enough to go through DBS. Some patients go through severe behavioral changes after surgery. Increases in anxiety and depression are not uncommon, as are mood swings and other psychiatric disorders. Some patients have reported developing addictive personalities, and some have reported an increase in suicidal thoughts. By this time I had fully developed my Warrior mentality, and I was excited and optimistic about the future. My depression had subsided a few months earlier, and I was ready for the huge change in my life. The doctors asked me questions about my state of mind, and they concluded that having DBS surgery was an acceptable risk for my psyche.

The next morning I had the **physical** test, testing my OFF neurological exam to an ON exam. I woke up in the surgery center and was given breakfast, then asked to take my non-Parkinson's medications only. This meant that I was off and would continue to be off until the nurse gave me a supervised dose of fast-acting Sinemet. The nurse turned on the video camera, then she first did a standard neurological exam (testing muscle tone and rigidity, finger tapping, foot stomping, getting up from a chair with my arms crossed). She then stopped the recording and gave me a fast-acting dose of Sinemet. This was like no Sinemet I had ever taken. Within 20 minutes I was on and halving dyskinesia. She then turned the camera back on and repeated the neurological exam. This time it was obviously very easy to do all of the tapping and stomping she wanted. She noted the my left side was a little slower, but that the Sinemet was working to its fullest. I would later learn that the nurse had noted a 350% improvement being on from being off, a really good indicator that DBS surgery would serve me well.

Next up, I had a visit with the **cardiologist** to see if my heart was healthy enough for surgery. This was as fully packed a visit that I'd ever had with a cardiologist. The cardiologist conducted an ECG, which had always come back normal, so I wasn't worried there. Then he took my blood pressure, which came out at a near-perfect 119/81, mainly thanks to the propranolol I was taking. Finally, the nurse took some blood, presumably to test for things like high cholesterol. I had been taking a statin for cholesterol, so I wasn't too worried about that. Heart disease ran in my family, and I told them about that, but that didn't seem much of a concern since the DBS unit wouldn't interfere or contraindicate any future diseases. However, because of my family medical history, they decided that the neurostimulator would be implanted under my right clavicle, in case I needed a pace maker in the future, which must be implanted by the heart on the left side.

Finally, I had a visit with an **anesthesiologist** to test if I was healthy enough for general anesthesia. This was the shortest of all of the meetings. I was asked if I had ever had anesthesia, and I said no and that this was my first surgery. We joked about how my first surgery would be a big one, and I remember saying something to the effect of "go big or go home," to which I received a blank stare and soon realized that this was likely an Americanism that hadn't yet reached the Netherlands. The doctor asked me if I had any problems with my lungs or if I had any respiratory issues. I had had pneumonia in December of 2014, but she said that wasn't an issue since it had obviously been taken care of. She cleared me for surgery then and there, wished me luck, and sent me on my way.

The whole evaluation experience made me feel like I was entering the space program. It felt good to know that despite having Parkinson's, I was otherwise in good health. I was ready for liftoff!

About a week after these evaluations, I received the call from the surgery center. The nurse had me in suspense as she went over all of my results, and only toward the end of the call did she tell me I had been accepted for DBS. I was so excited – I must have been the only person in history to be so happy to be having brain surgery! I was eager to meet the neurosurgeon and to start counting down the days.

Mid October – about one and half months before surgery

A few weeks later, I was back at the hospital to meet with the neurosurgeon who would be conducting the actual surgery. This meeting was not in the neurosurgery center, but in the polyclinic, which was an outpatient center full of offices in a different wing of the hospital. The neurosurgeon was tall with big, strong hands. He wasn't quite what I expected – he looked like he had just come back from holiday in France with his long hair and tanned skin. He was very confident of himself. He was a star and he knew it. Somehow this put me at ease.

I had read all about DBS already, so I knew about what would happen during most of the surgery. However, he went over every step of the process, as it's performed at this hospital. This is when I asked about their complication rate and their innovative surgery under general anesthesia, and this is when his confidence really shined. Now was also the time to select the hardware that would be used for the surgery, as well as have any additional questions answered.

The hospital had pioneered a new method of performing the surgery fully asleep, completely under general anesthesia, and their own research showed that this new method was just as effective in placing the leads correctly in my brain as being awake during the first half of surgery. This method was also safer, since they knew the exact spot for the lead, and there would be no guessing (taking the lead out and putting it back in while the patient tapped their fingers and did other tasks on the operating table). The neurosurgeon told me that the fully asleep surgery was technically not yet approved for general usage, but that I could chose to have that surgery since they were sure it would work. After hearing stories of people having nightmares for months from the drilling sound alone, I opted to be fully asleep during surgery.

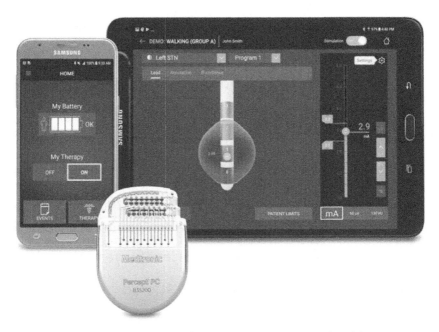

Figure 3: Medtronic's DBS system (© Medtronic)

Next, I had to choose the technology. The hospital worked with all major DBS manufacturers, and I could pick the hardware that I was most comfortable with or knowledgeable about. The hospital had had great success with the Boston Scientific hardware, which was not even approved in the United States at the time (it would be approved in May of the following year). I was leaning towards that hardware, since I had read great things about the new technology used to minimize side-effects by allowing the DBS programmer to steer the current away from areas in the brain that caused certain side effects. The Boston Scientific leads and probes were also much thinner than the Medtronic leads and wiring, which meant less brain damage as they entered the brain down to the target. Boston Scientific's remote control was also my favorite. The Medtronic control looked old fashioned (though they have since upgraded to an Android-based system), while Abbott used an iPod Touch as a controller. I felt uncomfortable with any device using a system as widely used as an iPod Touch or an Android tablet, since, as a developer myself, I know how easily those can be hacked and how much else can go wrong with them. The Boston Scientific remote looked sturdy, easy to use, and secure.

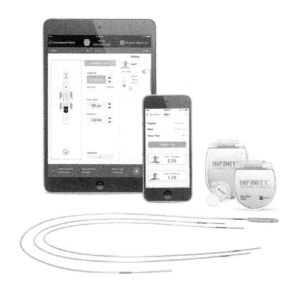

Figure 4: Abbott's Infinity DBS system (© Abbott)

Finally, I had to select the actual stimulator that would go into my chest. The stimulator was the control center of the system. It would house the battery – either a non-rechargeable battery that needed to be replaced every few years (as an out-patient procedure), or a rechargeable battery that needed recharging every 10 days or so but that would not need to be replaced for up to 25 years. I didn't like the sound of having my chest cut open every few years to change my battery. After all, I planned to live a long time, which could mean a dozen battery changes over my lifetime. The downside with the rechargeable battery is that I would need to remember to charge it. The neurosurgeon forgot that he was speaking with someone who grew up with technology – I had been recharging electronics like smartphones and mobile gaming systems for nearly my entire life. The doctor agreed and pointed to the fact that I was young, and I wouldn't need to worry about replacing the battery at least until my early 60s. With any luck, I would need to change it twice in my life! I knew I would never forget to "charge myself," so of course I chose the neurostimulator with the rechargeable battery pack.

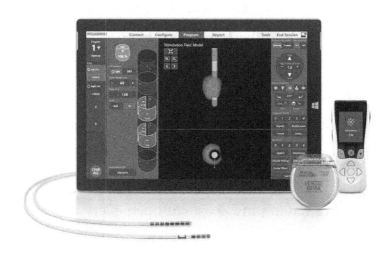

Figure 5: The Boston Scientific system I have installed. (© Boston Scientific)

One additional thing to consider when choosing a stimulator is a feature like MRI-mode, which allows you to safely get an MRI. MRI stands for Magnetic Resonance Imaging. Let me repeat the important part: magnetic. DBS puts wires in your head and a battery enclosed in steel casing in your chest. So the question is, will you be able to get MRIs once you have a DBS unit installed? The short answer is 'yes.' The longer answer is, it depends on the unit you have installed. If you are having DBS surgery after reading this book (presumably the case), then all three manufacturers of DBS equipment allow what is called MR Conditional, which is having an MRI with special precautions. Or as they are called by the manufacturers, "MRI, within approved parameters." This means that if you need to have a "head and extremities" or a "full body" MRI, this is possible with all of the newest equipment from all of the current manufacturers.

Next, the doctor was ready to schedule my surgery. I am not kidding when I tell you that he opened up a tiny black agenda paper notebook and asked, "how does December 6th sound? I said that was fine and he wrote the note in his agenda. I put it in my smartphone's joint calendar with Rosaline as "DBS!" He asked if I had any other questions, and then sent me on my way with a "see you on December 6th" and a pat on the back as Ros and I left his office.

Every pre-DBS meeting is different, and you may meet with the surgeon or with a nurse who will explain the procedure. Regardless of who you meet with, you will have an opportunity to ask questions. Here are some questions I would recommend asking during that time:

- **What is your complication rate?** Surgery is never perfect, and things do happen during surgery. Although complications are rare, knowing your hospital's complication rate will give you the confidence to move forward. Anything lower than around 3% is considered acceptable, but sometimes you're likely to hear complication rates as low as 0.06%!

- **Will DBS interfere with speech?** Sometimes DBS surgery does interfere with speech, and it can often be corrected with DBS units that have steering capability. DBS nurses can steer away from the areas causing the speech problems.

- **On which symptoms can I expect to see an improvement?** Remember, DBS only helps with motor symptoms that are alleviated with carbidopa/levodopa.

- **How long is the surgery?** A full bilateral DBS surgery with neurostimulator installed can take eight hours. It would take less time to operate on just one side, or just to implant the stimulator. Your surgery time could vary based on many factors.

- **How long is the recovery?** How long do you stay in the hospital afterward, and how long until the stitches come out? How long until you can go out in public? Some people go for walks the day after surgery, and some spend a week in bed like I did.

- **Will you perform the lead and stimulator surgeries together or at different times?** Depending on the hospital, the country, and the insurance you have, everything can be done at once or they can operate on one side at a time, and implant the neurostimulator at a different time.

- **How long until I get the unit turned on?** My DBS unit was turned on after 10 days, but I've heard some people waiting up to two months after surgery to get turned on.

- **How long is the initial programming?** When they turn you on, the initial programming could take four hours or it can take all day. This also greatly varies on your hospital and the skill of the DBS nurse.

- **How long does it usually take to fine-tune the DBS unit until it is properly adjusted?** Usually, the DBS nurse sets you up with settings that are commonly appropriate as initial starting points. It can take 6-12 months to get properly fine-tuned.

- **Will I need to get fine-tuned again every year?** As the disease progresses, you will likely need to be adjusted a bit each year.

- **In the future, is the stimulator upgradeable?** Often, new neurostimulators with new features (like rechargeability and longer battery life) can be installed over time.

Here are some questions you might want to ask about your particular health insurance. Your hospital may or may not have this information, but they can help you find it and you should know the answers before going forward with surgery.

- **How much does insurance cover?**
- **What is your insurance co-pay?**
- **Are all providers involved "in-network?"** One out-of-network team member on the surgery team can cost you a lot!

I couldn't believe my surgery was finally booked and it was really happening. I was very excited and began counting down the days. I would finally have some relief from this relentless illness. Rosaline and I invited my brother and my mom to stay with us during the week of surgery (my mom wanted to stay until the doctors turned the unit on so she stayed longer), We booked their flights from LA to Amsterdam, and we bought two blow-up mattresses. I would have my support system in place for the big day.

With just a couple of days until surgery, the hospital called me to confirm my surgery and told me what to take and what not to take of my medications. I should not take any Parkinson's meds, but I should take everything else.

December 5th – the day before surgery

I arrived at the hospital at around 1pm the day before the surgery. Rosaline and my family were all there, and I showed them the giant atrium and the Starbucks, telling them this is where they could wait while I underwent preparations for surgery and get their coffee the next morning. Then we went up to the DBS center on the 5th floor, and I was immediately assigned a room. For the next four hours, I was dressed in sanitary garments, had my vitals checked every hour, and introduced to all of the staff that would be assisting me during my time in the hospital. The DBS programming nurse came to check on me, asking me if I was ready. I was all smiles and told her I couldn't wait for morning. Around 5pm my neurosurgeon paid me a visit, and again asked how I was doing. Again, I told him I was excited and couldn't wait to "get drilled!" If the initial evaluations made it feel like I was joining the space program, this day felt like the day before my first solo space flight. My family stayed until closing time (around 6pm), and I told Rosaline to take my mom and brother to a have a nice dinner in the city.

The evening before surgery went as well as could be expected. I scrolled through Facebook on my phone, watched some CNN International on TV, had dinner, and tried going to sleep at around 9pm. My biggest concern about the whole process, to be completely frank, was not the drilling or the surgery, but the insertion of the catheter. I had also researched this part online, and read that it hurt immensely. Around 9pm, the night shift nurses came to see me and tried to talk me into inserting the catheter then and there. I fought back and asked them to insert it after I was under general anesthesia. Seeing that I wouldn't give in, they relented and said "let the surgery nurses handle this one." I was very proud of myself as I tried going to sleep. Of course, I would have little sleep that night because of my excitement, and perhaps a bit of anxiety too. I have a friend who is deathly afraid of surgeries, and I accompanied him once when he freaked out at the last second before they had a chance to operate on him. I was going into this with a positive attitude, so I knew the chances of this happening to me were low, but I still wondered what would happen if I suddenly snapped out of my Warrior attitude and yelled "I want my mommy!" Of course, my mother was there at the hospital, so it wouldn't have been an unreasonable request!

December 6th – the day of surgery.

I had been in and out of sleep all night, but I was up by 6am when the nurses came by to offer me breakfast. My family arrived at around 7am with lattes and croissants in hand. I had breakfast, took my heart medications and anti-anxiety meds, and then by 7:30am the nurses came by to take me into surgery. I said goodbye to my family as several nurses rolled me one floor down to surgery prep. We took the elevator and I was laughing the whole time; it seemed ridiculous at that moment for four nurses to be rolling the bed around the building, down the elevator, and into the surgery prep. Part of the feeling was the newness of it all, and part of it was genuine excitement.

I arrived in surgery prep and I met the anesthesiologist and surgery nurse as they prepped me for surgery, inserting my intravenous needle and wondering why I hadn't had a catheter put in yet. I explained my deal with the night shift nurses, and they laughed. When I was all prepped, my bed was retracted and then I was rolled into the operating room lying flat. The room was about 30 feet by 40 feet in size, and it was a buzz with about a dozen doctors and nurses prepping their respective areas. The surgeon appeared in my line of sight and asked me to verify my name and for what procedure I was there for. I said, "My name is Nick Pernisco, and I'm here to get some holes in my head!". Everyone laughed as they continued prepping.

After about ten minutes, and realizing I was calm and at ease with the whole situation, the anesthesiologist put his mask over my face and told me to breathe normally. The mask made breathing a little difficult, but by the time I was about to say something about it, I was out like a light.

December 6th, later that afternoon

I woke up disoriented in the recovery room. I was groggy and on painkillers, so I was unaware of what was going on around me. A nurse nearby seemed to be working on something near my bed, maybe checking my vitals or something similar. I wondered out loud, "where am I?" The nurse looked at me, adjusted my bed so that I could sit up a bit, and she said that I was in the recovery room. The surgery happened while I was asleep and so I asked, "did they do it?" I couldn't remember a single moment of the surgery – it's like I didn't exist during that time, I had no sense of consciousness, one finger snap and I jumped through time. So no wonder I wondered if it really happened, if I had really had deep brain stimulation surgery. I slowly began to awaken and noticed the bandages on my head and on my chest. I felt no pain, but I was becoming aware that the worst part was over. It really wasn't so bad.

The anesthesiologist paid a visit and tried asking me questions, but I was so out of it I could tell I was speaking gibberish. After about 30 minutes I was conscious enough to realize that I was not in the operating room, and the nurse took me back to my room upstairs on five, where my family was there waiting for me. I can't emphasize enough how much of a haze I was in since I had never had this feeling before. I still questioned whether it all really happened, and I kept asking my family, "did it really happen?" Rosaline and my brother were teasing me, because they said I was saying random things that didn't make any sense. Again, I'm emphasizing this part because the surgery is so scary for so many people, but I didn't remember a thing, and I was not in any pain coming off the anesthesia. The whole event was like a blip in my consciousness and I don't remember a thing.

What actually *did* happen is that I had a surgery of approximately eight hours. During this time, the doctors drilled their holes, imaged my brain, inserted a lead using a probe on each side of my brain (the holes are just beyond my hairline, and today they look like sawed-down devil horns), routed the wires under my scalp, down my neck under the skin, and into the tiny chest pocket below my right clavicle, where they connected the wires to my rechargeable stimulator. Later on I would count the number of wounds on my head and they turned out to be around ten: four small pricks from the stereotactic frame, two holes where the leads went in, two cuts where the doctors routed the cables under my scalp (they had to pull underneath the scalp from the hole to the side of my head), one cut in my neck to pull down the wires down to my chest, and one three-inch cut in my chest to fit the stimulator. The surgeon was able to cut the main holes in my head without me cutting my hair and without shaving the area, and hair still grows above those scars, but there is one part on the right side of my head where the wires were routed where the hair no longer grows. This is the permanent mark that serves as a reminder of the day I got my life back.

December 6th, later that night

When I was starting to wake up, my family got the call that surgery had gone well. They were already on their way to the hospital when they got the news. They stopped by the gift shop to buy me balloons (which I still have as a souvenir) and a cute pair of hospital slippers. When they got to the room, they realized I was still out of it. I was speaking non-sense. The nurse came to give me dinner – spaghetti and meatballs – and Rosaline lovingly cut it up and fed it to me. I realized I was still wearing the catheter, so I asked the nurse to take pity on me and to let me have two apple juices – everyone laughed, but I did get the extra apple juice! Visiting time was almost over, so I once again suggested to Rosaline that she should take my family to dine out, and maybe do some sightseeing. They all laughed at me – like they were going to sightsee while I was in the hospital. I'm sure that wasn't the stupidest thing I had said that day. Coming off of anesthesia is a unique and ultimately indescribable experience.

That night was a difficult night. I was no longer anxious or excited, but the painkillers were starting to wear off, and I could feel some discomfort on my wounds. The wounds on my head hurt a bit, and the one on my chest itched. It was difficult to get comfortable and get into a resting position, so I couldn't sleep much that night. I tried to sleep as much as I could, but probably only slept for about an hour intermittently. I watched television instead – probably some more CNN. The nurses came in every couple of hours to check on me, and they kept asking if I was ready to get some sleep. I told them I was ready but I couldn't get comfortable. They offered me more painkillers, but I only wanted the non-narcotic type, so they offered me two Paracetamols (essentially Tylenol) and after about an hour the medication started taking effect.

At around 6am, the night nurses came in and asked if it was ok to remove the catheter. I was a little apprehensive, but I knew it wouldn't be in there forever – sooner or later this time would arrive. I opened my surgical garment and looked away. She counted to three and I braced myself. On "three" I didn't feel anything, and after one second, it was over. My biggest fear had come and gone in a few seconds, and it didn't even hurt. Another moment of being scared for no reason.

December 7th – the morning after

I was eating breakfast at around 7am when my family arrived, this time with a soy latte and a croissant for me from Starbucks. I ate the hospital food then started on the Starbucks. I was wide awake and ready to go home. After eating breakfast (twice), I took my medications. I had to take them all because it would still be 12 days before they would turn on the stimulator, giving time for my wounds to heal and for the "honeymoon" period to be over (more on that later). I was visited by the neurosurgeon to see how I was doing. I gave him the thumbs up and told him I was in a little pain, but felt fine overall. After this, I gathered my things and I was discharged from the hospital. I was asked if I needed a wheelchair to the car area, but I declined; I came into the hospital on foot and I would leave on foot. The nurse offered me opioid-based painkillers, but I refused and said I would continue with the Paracetamol.

As soon as we got home, I could begin feeling the surgery incisions and so I took two Paracetamols and fell asleep. I continued taking two Paracetamols every 4-6 hours and going back to sleep. This continued for the next couple of days. I was so tired, my body had just undergone a huge stress, that I just could not keep my eyes open. I wasn't in pain, but rather I was in discomfort from the under-skin routing of the wires and from the chest implant. The hospital medications continued to wear off over the next few days, and the after-effects of surgery began. About two days after surgery, I developed black, swollen eyes. It looked like I had been in a terrible fight with a professional boxer. The continuous stream of pain medication kept the pain at bay, but I looked monstrous. Luckily the swelling and bruising began to disappear after about three or four days, right around the time that I was supposed to remove the bandages from my head and chest and take my first shower.

Figure 6: The swelling a couple of days after surgery.

The nurse told me I should shower and wash my hair about seven days after surgery. Until then I could shower but I would have to avoid getting the surgical wounds wet. I showered, but I was two afraid to wash my hair, since the top of my head was numb and I was afraid of pressing too hard. That numbness stayed with me for about nine months after surgery, but I eventually learned to wash my hair carefully.

Mid-December – the honeymoon period

Most people who undergo DBS note a period after surgery of about 7-10 days in which their brains are producing more dopamine and/or healing from the trauma of brain surgery (likely both are true, but no one knows for sure), and they are feeling great – almost Parkinson's-free. This is called the "honeymoon period" of DBS surgery, and it happens before the stimulator is even turned on. In fact, doctors want people to be healing and off their honeymoon period before they turn on the stimulator and start programming it to the patient's personal settings. I experienced this honeymoon period, during which time I noticed waking up feeling on, even before taking my Parkinson's drugs. I also noticed that my Parkinson's drugs had a faster and more pronounced effect during this time. My honeymoon period lasted about ten days, which was fine because I would have my stimulator turned on and have my first programming session two days later. Unfortunately, some people need to wait a month to have their stimulator turned on and programmed, but my wait was 12 days.

December 18th – time to get turned on

Long before my surgery ever happened, I was already scheduled to have my neurostimulator turned on and programmed just 12 days after surgery. I had read stories online about people needing to wait a month or two from surgery to powering on, and some people having to have the brain surgery first, then the neurostimulator implant happening at a later date, pushing the power-on day even further away. That is a long time to wait to feel the results of this surgery you just went through. Sometimes it has to do with the medical insurance a person has, and sometimes it has to do with patient safety. Luckily, my time from surgery to powering on was just 12 days. This may have to do with Dutch efficiency, who knows?

After 12 days of recovery, the swelling on my head and face were gone, but my head still felt raw and numb in some areas. Nonetheless, it was time to remove the stitches in my head – the DBS nurse used pliers to pick out about a dozen metal stitches, none of which hurt coming out. Then it was time for the big event. The DBS nurse connected me to a device that itself was connected to a laptop, and we were ready for the "lighting ceremony." My mom had stayed in Amsterdam for this moment, so there was some pressure to see improvement right away, though this isn't always the case. Sometimes it takes several programming sessions to get the settings right, and most of the time the patient needs to turn up their own stimulator bit by bit at home using the remote control.

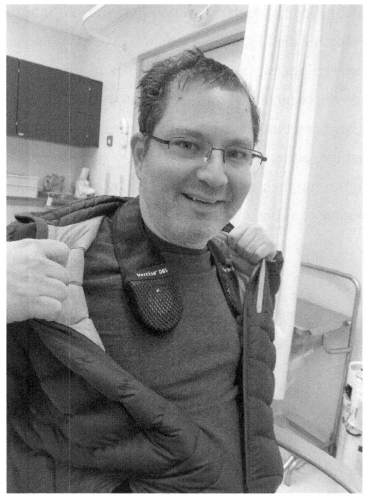

Figure 7: Turned on and charging myself for the first time.

After I was connected to the laptop, we spent the next three hours testing the various electrodes on the leads, making sure everything was working properly and discovering what side effects, if any, would occur as we tried different electrode settings. Each side's lead has eight electrodes that can be used at different voltages to get a particular result. The bottom-most electrode is a full electrode, the next two up the wire are actually 3-in-1 electrodes that allow for directing the current towards one side and avoiding other sides, and finally the top-most electrode is a full electrode again. Using her laptop, the DBS nurse could, in theory, try thousands of combinations to see what works best for my personal situation. We started by choosing the main electrode – the electrode at which I would have the most effect, which for me was the second electrode from the bottom (a 3-in-1 electrode). My initial setting was on that electrode on all three sides, but at different voltages on each side. This was a good starting point, and although it would need to be fine-tuned later, I already had improvement in my hands and feet.

After the initial programming had been done, I was given the neurostimulator remote, which would allow me a certain range to turn up or down the voltage on my unit, but only on the current program – I couldn't change electrodes or turn myself up to 11. I could make minor adjustments to voltage only. I had started low and noticed I was having to "turn myself up" every few weeks or so. I was told that this was normal and I could continue going higher until I was stable and didn't experience dyskinesia. I was officially a cyborg – part human, part machine – and I even had the charger and remote to prove it!

Figure 8: Celebrating the night of the "lighting ceremony." If you look closely, you can see the small wounds from the stereotactic frame on my forehead.

The changes in my symptoms from before DBS surgery to after DBS surgery were immense. After just a few weeks, I was able to use less and less of my Parkinson's medication (lowering my dosage as I increased my voltage). I could turn in bed and I could walk quickly without dyskinesias (sometimes Rosaline would need to tell me "slow down!"), I could even run, an ability that I used in order to catch departing trams a couple of times, making it just in time like in the movies. I could type again, only an average of 30 words per minute, a pittance compared to my one-time record of 100 words per minute, but much better than my pre-DBS 7-10 words per minute. Those I met after surgery had no idea I even had Parkinson's, telling me they were surprised – until I showed them the scars.

Even though I was no longer showing motor symptoms of the disease, the result was not all roses, however. I still had Parkinson's, the disease was still progressing, and DBS only worked to better hide motor, not non-motor symptoms. Fatigue persisted, as did the occasional cloudy feeling in my brain. For about a year after surgery, I also had problems with short term memory as well as recall ("Who was that actor from that movie? It's on the tip of my tongue.") Years later however, it seems that the memory and recall are much better, and the fatigue has subsided thanks to exercise.

Everyone will have a different experience, since everyone with Parkinson's has different symptoms to begin with, but my overall experience is not likely to be too different than yours. Around the world, doctors use the same techniques, the same hardware, and the same evaluation procedures. If this is something you feel you might benefit from, it's worth discussing with your movement disorder specialist. It might change your life like it changed mine, and it might just give you a break from all of the difficult motor symptoms we experience.

Now that I have been given another chance at life, I haven't taken it for granted, even during the great quarantine of 2020. I was determined to use my time wisely, so I wrote this book, earned a professional certificate in data science from IBM, and began working on a rock album. In many ways, DBS absolutely met my expectations. I feel like a new person. But in other ways, the surgery left me mourning for symptoms that haven't gone away, and have thus become my main Parkinson's symptoms – my non-motor symptoms. I'd like to tell you all about my new life in hopes you might learn from my experience and be inspired by your own new potential after DBS.

It's a New Life, It's a New Day

"When I let go of what I am, I become what I might be." —Lao Tzu

From the moment I took off my bandages from surgery, I began to feel like a new person. Something had changed in me, and it wasn't just the holes in my head or the device in my chest. I felt like life had regenerated itself within me – a new spring had come, so to speak. I was off the high doses of carbidopa/levodopa, and my body was not only surviving, but it was thriving. My body had always been quick to regenerate itself – my hair and nails grow quickly, and within a month or so after surgery, the wound on my chest was barely noticeable.

I still had Parkinson's, that wasn't going away without a cure, but I could finally do things I couldn't do before. I could type again – the original Parkinson's Warrior book was dictated, but this book was proudly typed – and I could do a whole host of things I couldn't do before. I could walk. And not just a better walk than my Parkinson's walk when I was unmedicated, but I could actually keep up with and even walk more quickly than Rosaline. I've never been much of a runner, but now I could run in sprints as well. Symptoms that were immediately present upon diagnosis were suddenly gone – tremor, rigidity, and slow movements, had all disappeared. I could move my left arm freely and without giving it a second thought.

One of the things I had promised Rosaline was that we would go out for walks more often. She would usually walk in the park alone because I couldn't keep up and would need to quit halfway in. Now I could walk in the park and everywhere around Amsterdam for hours on end. My stamina was incredible. In 2019, we decided we would take advantage of all we had missed the previous year while I awaited surgery. We bought a museum pass that let us into almost every museum in Holland, we visited Rosaline's friends who lived on the opposite side of the city (walking, taking the tram, then walking some more), and we could finally go to restaurants with sloppy food without my fear of looking silly when eating with my hands.

Most of all, the fear of judgement was gone, not just in the Netherlands, but everywhere I visited. I no longer had to worry about being watched judgmentally by people on the street. I never had to worry about being asked if I was alright by a passerby or a waiter. I could finally meet new people without them worrying that I had a problem. From this point forward, no one would know I had Parkinson's unless I told them. But since Parkinson's was a part of my life, and a part that was still with me, I would not hide it if aspects of it peeked through. I looked Parkinson's-free, but I still had the fatigue, and at least at first, logic and memory problems. I also had built a reputation as a Parkinson's advocate and didn't want to lose that part of who I was. I was still in the fight to raise awareness and to help others struggling with the illness.

Despite using my new energies to continue helping the Parkinson's community, I did have a rare opportunity for a do-over in certain parts of my life. After fighting particularly difficult physical symptoms for almost a decade, I had what many people ask for when their life is at its lowest – a second chance; a new lease on life. I could rethink certain things I had already taken for granted as "this is how my life is going to be." Did I want to continue teaching, for example, or could I do something else more exciting with the new time I had been given? Could I go back to school, switch careers, or do something completely new like join an amateur sports league?

I tested my new "super powers." I gave speeches about Parkinson's, and presented the Parkinson's Warrior book at local book shops in Amsterdam. I was not only continuing my commitment to the Parkinson's community, but I was testing my limits. Being on stage still did give me fatigue, and I knew I would need to keep my body fit if I were to continue doing this. In the first year, even giving a talk for 10 or 15 minutes would leave me wiped out for the rest of the day. This was strange since I had no problems walking for hours. I suspected it was nerves and the stress of being "on" (like an actor on stage, not like a person with Parkinson's taking Sinemet), and not being fit enough to handle it physically. So speaking to groups of people or using my voice loudly would not be something I'd be doing a lot of for the time being. This meant that my dream of giving a TED talk about the power of positivity on disease progression would need to wait.

I recommitted myself to the app I had created for the Parkinson's community, Parkinson's LifeKit. I upped my efforts and even merged the content from Parkinson's Warrior with the functionality of the app. I created new features that customers loved, but I felt that the more intense work of developing and selling a complex app left me depleted for the rest of the day, and at times for two or more days. It took a lot of mental effort to not only research and design a solution to a problem, but to also implement it in the form of an app and that customers would find useful. This is when I realized that growing my company even larger by engaging with investors was not going to work during this time. It would simply be too stressful. So my dream of building a large company creating great products for people with Parkinson's would need to be put on hold.

I continued working on the app on my own and at my own pace. I continued talking to people in the Parkinson's community – patients, doctors, physical therapists – and continued making the app better for those who wanted to track changes in their symptoms on a daily basis. When I needed a rest, I would take a couple of days off then come back to it when I was no longer tired. I could continue building my software company, but on my own terms. I didn't have a boss or investors breathing down my neck giving me deadlines or asking me to work on the weekend. I could make my own hours and work when I felt I could. This turned out to be a dream scenario.

I was such an avid tennis player before Parkinson's that I was eager to jump on the court to try out my new skills. It turned out that tennis was another activity that would leave me disappointed at first. Once again, non-motor symptoms like fatigue and balance issues had me unable to keep up with opponents on the court. If I could reach the ball, I could hit it hard and with great accuracy, but I still couldn't play for more than an hour. I was also afraid that anytime I would step backwards to reach for a lob would end with me failing flat on my back.

This entire process of rediscovery was like a second round of denial and acceptance. I had already accepted I had Parkinson's and had adjusted to my new life as someone who had a limiting disease, but now it felt like I was going through this entire process again. I was told that fitness could help relieve a lot of these limitations, but it was still a disappointment that I couldn't do the things I so wanted to right at the start. My DBS nurse had said that many patients experienced this sense of unmet expectations. That patients had expected to at least feel like they had been cured or at minimum, have most of their function back, but that the surgery, while a success in many ways, still didn't provide full remission like they had wanted.

Looking at it from a broader perspective, I couldn't look at any of my experience as a failure. I had regained my mobility. I could resume my life in many ways. I had this new super power, a gift given to me so that I could improve my life. When I felt myself getting depressed about not returning to complete normalcy, I stopped myself. I took a good look in the mirror and said to myself, "You've been given a gift. Maybe you're not meant to give speeches or be a startup CEO, or to compete against Roger Federer." The truth is, I never wanted these things. I realized that speaking to a group of people with Parkinson's was enough for me. It always had been. Making an app in my spare time was really what I wanted to do, not become the next Bill Gates. And it used to be that as long as I could play tennis for an hour with a friend, I would have a good time.

I was trying to be more than I needed to be. I had spent much of my life trying to fit a square peg in a round hole, and this was more of the same. That day I came to this realization – I hadn't been given this second opportunity in order to be someone else, to be someone who only existed in my imagination, but to be the best *me* I could be. This realization changed my outlook on my new life. I wasn't supposed to do as much as possible and be all I could be, like the old Army slogan that makes you dream big. My purpose was to be who I already was, only better.

Equipped with this new realization, I asked myself the eternal question, 'What was my end game?" In other words, what did I want out of my life? Not, "what task should I complete or what role or job should I have," but rather, what was my *purpose* in life? After I had died, I didn't want to be remembered as great speaker, or a great programmer, or an amazing tennis player. I wanted to be remembered for my advocacy and activism in the community. I wanted to be remembered as a good husband, as a good brother, as a good uncle, as a good friend. I wanted to be remembered as someone who proudly served college students by teaching them to be mindful adults. That's what deep brain stimulation surgery had given me; the possibility to be the best version of who I already was. That's what, as it turned out, brought the most joy to my life.

I was meant to be an even better husband – someone who could go out to dine and dance, where this wasn't possible before. Rosaline had been by my side through everything. She was with me through diagnosis and depression, and through DBS and recovery. She means everything, I didn't want to jeopardize that for anything, let alone for something that had no purpose or that brought me no real meaning to my life. I was also meant to be a better uncle, traveling to visit my nieces more often and offer them guidance based on my own life experiences. My older nieces were all grown up, but my youngest niece could still use her uncle's guidance. These are things I was already doing and had wished I could do more of. Now I could.

There was also no need to find a new career – I enjoyed teaching young people, and my heart fluttered whenever I received an e-mail from a student who had been accepted to the school of their choice, and had succeeded to go off to do something great. I loved getting e-mails and Facebook messages from people with Parkinson's telling me that my app or my book or my website or my Facebook group had helped them. I didn't need to be perfect, or speak perfectly, or write code perfectly – I just needed to be a better me. In the past I would get depressed hearing about Rosaline's business adventures, wishing that it was me launching a new company and living the startup life. I lamented the fact that I didn't have any of that in my life. But now I was thrilled to hear about Rosaline's work day. Instead of feeling jealous or angry that I was not able to work a high stress, full-time job, I listened to her stories and shared in her excitement about her work. This has given me a newly-found appreciation for the work she does and inspires me immensely.

Now when I think about the things I do in my life, I don't focus on individual tasks – researching, programming, advertising – instead choosing to focus on meaning and purpose. What is the overall purpose of what I'm doing? I used to love filmmaking, and I knew that was something that Parkinson's had taken away from me. So with renewed purpose, in my last few months in Amsterdam, I connected with the local filmmaking community, read off my resume, and asked to be involved in any way possible. I didn't care what I did on set, that was the task. The overall purpose was to get back to filmmaking, which I loved so much. I ended up being the film's lighting person – a job I had done for my own work but never on someone else's work. I didn't care. I was on the set, working with the producer and director, meeting new people, and helping them achieve their vision.

The film came out great, and after it was done, several people asked me if I was working on anything new. I told them that I had written a short script about a man with Parkinson's who is stuck in his apartment. Almost immediately, I was encouraged to shoot it and was offered free help in every department. The community was ready to work with me on my passion project. I realized that if I shared my passions and ideas with others, I could get people to participate in my short film. Here I was, being myself, doing what I already loved, but doing it with purpose. That ultimately attracted people to my work, that I was doing it because it gave my life purpose, not because I would become rich or famous.

Everything I do now I do with the end goal in mind. This has given me a longer-term view on life. For every task I undertake, I think about my "end game." What is the ultimate purpose of what I'm doing? If it doesn't contribute to an important purpose in my life, then I don't do it. This doesn't mean that I should only do things that are serious and long-term – I want to have fun like everyone else, to have moments in which I don't take myself seriously, to allow myself to be silly. But purpose can come even from those activities. An example for me is playing pinball. I used to play local tournaments in the Seattle area with the hopes of rising in the international rankings and becoming a top-1,000 player (my highest rating was about 2,000 at one point). Although I still played some tournaments, my focus turned to playing for fun rather than playing for ranking.

Getting together with friends for some pinball adds more meaning to my life than being a top-1000 player. After all, what would that get me? No one would treat me differently, and I'd be the same person. There is no purpose or meaning for me in being a top player.

My experience with DBS has also taught me to separate the idea of doing things for fun rather than for mastery. In June of 2019, I visited friends and family in the United States, and made time to attend the Northwest Pinball and Arcade Show. This was an opportunity to reconnect with friends in the pinball community, and to find out if Deep Brain Stimulation had improved my pinball skills. I had been a part of the pinball community since about 2015, and being a part of it held meaning for me. Although I was no longer competing to be a top player, it meant a lot to me to show some improvement at the game (players often compete one-against-one, and it's no fun constantly losing, even if it is for fun). I played in all of the tournaments at the show, and I ended up doing better than I had two years before. The DBS had done its job of unfreezing my hands and body so that I could hit the flippers at the right time, and even nudge the machine when needed. Here was another instance in which DBS was helping me be a better me. That was a fun weekend catching up with friends and playing pinball for nearly 72 hours straight (with a few naps in between!).

The change I underwent after DBS was all about attitude. Just like any good Parkinson's Warrior, I adjusted my attitude to the situation. Eventually I decided that DBS had met my expectations, even if it wasn't a complete cure. Instead of being down and also getting everyone around me down, I chose to be happy about everything I had gained. Life has been fantastic ever since.

Takeaways

In telling you my story about having had DBS surgery, there are some ideas, notions, and thoughts that I hope you walk away with. After all, DBS surgery worked well for me, and it has worked well for tens of thousands of people with Parkinson's around the world. The reason I believe that more people don't have the surgery is because they are afraid, because, let's be fair, *brain surgery* sounds intimidating and scary, but also because people don't fully understand the surgery and its benefits. I hope this book has shown you that DBS doesn't have to be scary, and I hope I have given you enough information to discuss DBS with your movement disorder specialist, and then to decide if DBS is the right solution for you.

So without further ado, let's take a look at some takeaways I hope you have gleaned from this book.

Don't let fear hold you back

Fear is a complex beast, and it can quickly overcome you and have its way with your emotions. Don't let fear get the best of you, and the best way to do this is to confront fear head-on. Learn as much as possible about the surgery. Go to YouTube and type "Deep brain stimulation" into the search box. Learn as much about the surgery as possible. Watch a surgery being performed. Listen to the experts. Listen to the patients. The more you know, the more comfortable you'll feel about it, and the less fear you will have about the process.

One caveat to watching YouTube videos about DBS, is that there are a lot of organizations that only show you the good parts. Medtronic, Boston Scientific, and Abbott aren't going to show you the scariest parts of the surgery, since they are trying to get you sold on having the surgery and to use their hardware. Instead, watch videos that patients or doctors have posted themselves. Videos posted by hospitals are good for getting an idea for how they perform the surgery, but remember that many hospitals are also trying to sell you on having the surgery at their location. I would recommend you watch as many videos as you can, because the more you know the less scary the surgery becomes.

Don't think life is over, it could just be beginning

Many of us with Parkinson's have thought, at one time or another, that our lives were over. These feelings are understandable when you have just been diagnosed, or when you are having a particularly bad day. It's important to constantly track your symptoms to find out if your feeling bad is just an off day or if it is a part of a larger trend you should be concerned about. This is why I created Parkinson's LifeKit, an app that allows you to track your symptoms daily. It's being used by thousands of people each day to track their physical, psychological, and cognitive symptoms.

Even if you are experiencing a rapid decline in your symptoms, your life is not over. Not only could you improve your body by developing a fitness routine and eating well, but there are also advanced therapies like DBS that can help mask the symptoms and give you your life back in very important ways. If DBS is something you have just started considering, remember that there are other therapies you can try while you are still deciding, like dopamine agonists and devices like the Duopa pump, which can deliver a constant dose of carbidopa/levodopa through an implanted pump.

Don't lose your sense of humor

Life can be difficult as it is, let alone when you're dealing with a debilitating illness. It can be easy to become jaded and become bitter at society, care takers, and loved ones. When something goes wrong, it's easy to get angry and place blame all around you. It can also become very easy to let every little thing seem like the end of the world. The suffering is real, but it doesn't always need to be. There are times when small negative events happen and, depending on your attitude, you can respond with despair, or you can respond with a laugh. Sometimes we don't have control of a situation, but we always have control over how we react to it.

An example I like to use is when cooking or eating dinner. I'm not the best cook, so Rosaline and I recently decided to buy these weekly meal boxes that come with recipes and all the ingredients for three full meals. I call it cooking by numbers, since the recipes feel like those childhood coloring books – just follow directions and voila, a meal will get made. Except I'm sure they don't expect a person with Parkinson's as the sous-chef! At first, I would get frustrated by the number of steps, the stress of needing to keep several pots and pans going at once, keeping track of how long is left to cook a particular item, and whether some meat is fully cooked or not. I couldn't do most of these things pre-Parkinson's, so now it's even worse. Sometimes I feel like we're in the I Love Lucy episode where Lucy gets a job in a candy factory. As a person with Parkinson's, I could either get discouraged every time the kale pieces end up on the ground and the seared pork gets over-cooked, or I can just have a laugh at the thought of me making these complex meals. Of course there will be disastrous results from time to time, but I've come to expect it and can laugh at myself instead of getting frustrated. It's all in how you approach it.

Don't take yourself too seriously

We have Parkinson's. We are occasionally in need of medical attention. We take some serious medications. It is still ok to make light of some situations. I am the first to excuse myself for not remembering something by claiming, "well, I have holes in my head, literally!" Or I'm pushing the cart in the market and I say out loud, "I know I can, I know I can," and then when I reach the register I exclaim, "I knew I could!" We have an illness, but we still need to have fun in life with what we have. Again, it's a matter of attitude. In life, as with so many other things, you get out of it whatever you put into it.

Never give up hope

There'll be days you'll feel terrible, and there'll be days you'll feel great. Sometimes, you'll have a general feeling of malaise for weeks on end. No matter what therapy you're using, there will be times you will feel like giving up. There is just no way around that. Even with DBS, I am still fully aware that I have Parkinson's and I know that my symptoms will worsen over time. It's during this time that you should not give in to the illness, and keep moving forward with hope.

New treatments are being released all of the time. DBS is a treatment that has improved over the last two decades, and it offers new hope for those needing relief. Believe me when I say that I had given up hope many times, but after a day of thinking rationally about my options, I knew DBS would give me that extra time needed for a cure to be developed. Again, this is hope. I can now live a much better life while I wait for a cure. With DBS, I will always be better off than without it, but DBS too will one day not be enough to keep me going. I still hope for a cure like everyone else.

Be strong

No matter how hard the disease hits you, you must be strong and hit back even harder. At the beginning of our Parkinson's journey, this means taking the right supplements, getting enough exercise, eating the right foods, and taking the right medications, not to mention getting the right medical help. Things get harder as the disease progresses, and you need to fight back even stronger. This can take its toll, but you must be strong. You don't have any other options but to be strong. If you're having a bad mental day, you must push through it by finding something to help you change your mood. If you're having a bad physical day, get up and walk to the kitchen or around the block. And if it's time to consider an advanced treatment to live a better life, consider it. It takes mental toughness to decide to go through with DBS, but you must find your strength if you want to reap the rewards.

Be positive

It's difficult being positive if you're surrounded by negativity. Put yourself in a place (real or imagined) in which you can be positive with your thoughts. Many times, your body follows your thoughts. If you are negative and see the world through sad, bitter eyes, your body will react accordingly and you'll feel terrible. However, if you break free of your low mood and "get to your happy place," your body will react in positive ways that may be unexpected.

The notion of being positive or being negative often comes from the state of your mental health. Depression and anxiety contribute to an overall negative feeling about life and the world. Don't forget to tend to your mental health. Remember, DBS only helps with motor symptoms, so any depression or anxiety that you had before surgery are still likely to be there after surgery. Make a solid plan to receive treatment for your mental health and you will be rewarded with more positivity in your life.

Be a Warrior

Of course you need to be a Parkinson's Warrior above all. Having Parkinson's is the biggest battle of your life, and the battle against it must be fought as such. The doctor who Parkinson's is named after called it "The Shaking Palsy," but I go further and call it "The Ballsy Palsy." That's because Parkinson's is the ultimate unwanted guest – it arrives uninvited, never wants to leave, and expects to take over as soon as possible. We need to fight it every day, even if it's just a little bit. Everyone will do this bit differently, since we are all affected differently, but we must do it.

There is a quote that I love that sums up the Parkinson's Warrior mentality:

"Fate whispers to the warrior, 'You cannot withstand the storm.' The warrior whispers back, 'I am the storm."

The quote's author is unknown, but wow, how powerful (and empowering) it is! Your fate is in your hands. You are the one in control of your life, not Parkinson's.

Conclusion

To sum up my experience with Deep Brain Stimulation, I say this: many people who I know who have had DBS, have said it is not a perfect solution for all their symptoms. With that said, they have all told me that they would have the surgery again in a heartbeat. I will go one step further: if I had to have DBS surgery every year for the rest of my life in order to keep the benefits I've had with DBS, I would do it in a heartbeat. Again, my case has admittedly had great results. I had it done at an ideal age, at an ideal stage of the disease, and I have had the ideal attitude for someone about to go through brain surgery. Your mileage may vary.

Of course, that's not to say you may not also have a great outcome. I have met many people with recent DBS surgeries and they are happy with it, as are those people with DBS units that had been implanted 10 or 15 years ago. Their current motor symptoms are a lot less than they would have been without DBS. It's that stimulation that will continue working the rest of your life, always lessening your symptoms. This is what I'm most happy about. I hope to never again feel the way I did before surgery. I know that my body can wear down over time, and that my disease will still progress, and that even with DBS I may end up having really bad on/off periods, and I may need to take medication again, and I may have dyskinesia again, but knowing that it could be even worse without the DBS installed makes me hopeful, and makes me think that if you believe in miracles, DBS is as close to one a person with Parkinson's can have.

I sing the praises of DBS, but I understand that deciding on whether to have the surgery or not is a very personal choice. If you are still early in the disease, you may be getting benefits from exercise, supplements, and medication. But if you have had the disease for a decade or longer, and you are hoping for your life to be so much more than it currently is, you may start thinking of advanced therapies, which include DBS. You could also be in the older group of people with Parkinson's, where you are in your 60s, 70s, or 80s, and you are just happy to be alive, exercising, and spending the time you have left enjoying every moment. These are some of the profiles of people with Parkinson's I've met. But by no means are they meant to be reduced to a stereotype. I know of people in their 20s, 30s, 40s, 50s, 60s, and 70s who have all had DBS surgery.

For me, it was having surgery at age 40. After struggling for some time, I knew it was time to try something drastic. Now I've been given another chance to do something purposeful with my life, and to use the second half of it doing things for my friends, my family, and my fellow Parkinson's Warriors. I've been given a second chance at life!

Resources

Parkinson's Warrior Facebook Group:
https://www.facebook.com/groups/657589511388428/

Parkinson's Disease Research video:
https://www.youtube.com/watch?v=9phXvB077Dw

Abbott DBS
https://www.neuromodulation.abbott/us/en/products/dbs-therapy-movement-disorders.html

Boston Scientific DBS
https://www.bostonscientific.com/en-EU/products/deep-brain-stimulation-systems.html

Medtronic DBS

https://www.medtronic.com/us-en/healthcare-professionals/therapies-procedures/neurological/deep-brain-stimulation.html

The Michael J. Fox Foundation: DBS

https://www.michaeljfox.org/news/deep-brain-stimulation

American Parkinson's Disease Association: DBS

https://www.apdaparkinson.org/what-is-parkinsons/treatment-medication/deep-brain-stimulation/

Parkinson's Foundation: DBS

https://www.parkinson.org/Understanding-Parkinsons/Treatment/Surgical-Treatment-Options/Deep-Brain-Stimulation

Parkinson's UK: DBS

https://www.parkinsons.org.uk/information-and-support/deep-brain-stimulation

References

Wikipedia:
https://en.wikipedia.org/wiki/Deep_brain_stimulation

"Deep Brain Stimulation for Parkinson's Disease"
https://www.sciencedirect.com/science/article/abs/pii/S0959438803001739

"Pallidal versus Subthalamic Deep-Brain Stimulation for Parkinson's Disease"
https://www.nejm.org/doi/pdf/10.1056/nejmoa0907083

"A Randomized Trial of Deep-Brain Stimulation for Parkinson's Disease"
https://www.nejm.org/doi/full/10.1056/nejmoa060281

About the author

In 2011 Nick Pernisco was diagnosed with Parkinson's disease – a disease typically affecting those over 60 – at age 33. After enduring a prolonged period of depression, he found a way to move beyond his grief and fight back by taking control of his disease and is using his experiences to educate and advocate for others. From this mission, he has created several resources to help others living with Parkinson's.

The Parkinson's LifeKit app is used by thousands of people with Parkinson's around the world to help them take control by tracking symptoms and fitness, managing medication, and more accurately reporting on their condition to their doctors.

Parkinson's Warrior was launched as a news and resource website in 2018, and has since become a platform to publish books and other media, and has spurred a Facebook support group, all of which serve as guides to adopting a Warrior Mindset to take control of each day and each battle in pursuit of an improved quality of life.

In 2018, after years of disease progression and failing medications, Nick decided to have Deep Brain Stimulation surgery to help ease the symptoms of the disease. Since then, Nick's quality of life has improved significantly. Though not a cure, it has changed the way he lives his life and has changed his views on life in general.

In addition to being an author, app developer, and advocate, Nick is also a media studies professor. He has lived in Buenos Aires, Los Angeles, Seattle, Amsterdam, and most recently, New York, where he lives with his wife and two cats.

Made in the USA
Monee, IL
17 October 2020